PHYSICAL CHEMISTRY OF
COLLOIDS AND MACROMOLECULES

INTERNATIONAL UNION OF PURE AND APPLIED CHEMISTRY
Physical Chemistry and Macromolecular Divisions
in conjunction with
UPPSALA UNIVERSITY
SWEDISH CHEMICAL SOCIETY
ROYAL SWEDISH ACADEMY OF SCIENCES
ROYAL SWEDISH ACADEMY OF ENGINEERING SCIENCES

Honorary President HERMAN F. MARK (USA)
President of the Symposium BENGT RÅNBY (SWEDEN)

International Honorary Committee

S. Brohult, Chairman (Sweden)　　　M. H. Holmdahl (Sweden)
N. Gralén, Co-Chairman (Sweden)　　H. P. Lundgren (USA)
S. Arrhenius (Sweden)　　　　　　　H. Mosimann (Switzerland)
A. Banderet (France)　　　　　　　 R. Signer (Switzerland)
J. Burton Nichols (USA)　　　　　　S. Ulfsparre (Sweden)
P. Ekwall (Sweden)　　　　　　　　 J. W. Williams (USA)
G. Hambraeus (Sweden)

National Program Committee

B. Rånby, Chairman (Stockholm)　　 I. Jullander (Stockholm)
M. Almgren (Uppsala)　　　　　　　 K. O. Pedersen (Uppsala)
P. Flodin (Göteborg)　　　　　　　 P. Stenius (Stockholm)

Local Organizing Committee

B. Enoksson, Chairman　　　　　　　Ingrid Fagerström, Secretary
I. Moring, Co-Chairman　　　　　　 Edmund Schjånberg, Accountant
W. Brown

Organized by
Uppsala University, Sweden
Supported by
National Swedish Board for Technical Development (STU)
Swedish Natural Science Research Council (NFR)
Swedish Council for Planning and Coordination of Research (FRN)
and private foundations

INTERNATIONAL UNION OF PURE AND APPLIED CHEMISTRY

PHYSICAL CHEMISTRY OF COLLOIDS AND MACROMOLECULES

Edited by

BENGT RÅNBY

Royal Institute of Technology, Stockholm, Sweden

The(odor) Svedberg

Proceedings of the International
Symposium on Physical Chemistry of
Colloids and Macromolecules to celebrate the
100th anniversary of the birth of Professor
The(odor) Svedberg on 30 August 1884,
a Pioneer in Colloid and
Macromolecular Research
Uppsala University, Sweden:
held on 22−24 August 1984

BLACKWELL SCIENTIFIC PUBLICATIONS
OXFORD LONDON EDINBURGH
BOSTON PALO ALTO MELBOURNE

© 1987 International Union of Pure and Applied
Chemistry and published for them by
Blackwell Scientific Publications
Editorial offices:
Osney Mead, Oxford OX2 0EL
8 John Street, London WC1N 2ES
23 Ainslie Place, Edinburgh EH3 6AJ
52 Beacon Street, Boston, Massachusetts
   02108, USA
667 Lytton Avenue, Palo Alto, California
   94301, USA
107 Barry Street, Carlton, Victoria 3053, Australia

First published 1987

Printed and bound in Great Britain
at the Alden Press, Oxford

DISTRIBUTORS

USA and Canada
   Blackwell Scientific Publications Inc
   PO Box 50009, Palo Alto, California 94303

Australia
   Blackwell Scientific Publications
   (Australia) Pty Ltd
   107 Barry Street, Carlton, Victoria 3053

British Library
Cataloguing in Publication Data

Physical chemistry of colloids and
   macromolecules.
   1. Colloids   2. Polymers and polymerization
   I. Ranby, Bengt
   541.3'45   QD549

ISBN 0-632-01536-5

Library of Congress
Cataloging-in-Publication Data

International Symposium on Physical Chemistry
   of Colloids and Macromolecules (1984 : Uppsala
   University)
   Physical chemistry of colloids and macromolecules.

   At head of title: International Union of Pure and
Applied Chemistry.
   1. Colloids—Congresses.   2. Macromolecules—
Congresses.   3. Svedberg, Theodor, b. 1884.
4. Chemists—Sweden—Biography.   I. Svedberg,
Theodor, b. 1884.   II. Ranby, Bengt G.   III.
International Union of Pure and Applied Chemistry.
IV. Title.
QD549.I636   1984   541.3'451 86—26903

ISBN 0-632-01536-5

Table of Contents

Preface

The IUPAC Symposium on the "Physical Chemistry of Colloids and Macromolecules" (the Svedberg Symposium) was held on August 22-24, 1984, at Uppsala University, Sweden, to celebrate the 100th anniversary of the birth of Professor The(odor) Svedberg, born on August 30, 1884. The Symposium was attended by 220 scientists, of whom 100 were from abroad, representing 17 countries. In addition, 40 Swedish senior high school students (20 boys and 20 girls) attended by invitation on special fellowships. These students were selected by their teachers to be the best chemistry student of his/her high school. Most of the students had previously been Berzelius Fellows at the annual Berzelius Days for high school students, arranged by the Swedish Chemical Society.

The Symposium programme offered a total of 49 presentations, of which 10 were introductions and other short addresses, 6 were retrospective lectures on Svedberg's contributions to science, 6 were lectures on colloid and surface science, 7 on sedimentation analysis of proteins and polysaccharides, and 20 on macromolecular solutions and gels. These areas of research, together with isotope and radiation studies, represent the main fields of activity of Svedberg and his students and associates.

The Symposium Volume contains the main addresses and the invited lectures. The retrospective lectures describe the intense and successful work of a great scientist in a wide area. We can follow how Svedberg's main interest in research gradually shifted from inorganic colloids to proteins and other macromolecules, and during the last period, after his retirement as professor of physical chemistry at 65, mainly to nuclear and radiation research. Svedberg's extensive activity in research formed the basis for three new departments of Uppsala University: Physical Chemistry, Biochemistry, and Nuclear Chemistry.

The main lectures invited describe current research in the "Svedberg areas". They reflect the importance of Svedberg's pioneering research on colloids and macromolecules. The development of ultracentrifugal sedimentation, and other preparative and analytical methods for protein research which followed, has made Uppsala a leading research centre in the field of protein analysis.

Polymer research started in the 1920's when the concept "macromolecules" was proposed and after a decade of studies and discussions generally accepted. In his introductory lecture, "Polymers in the 1920's", Professor Herman F. Mark gave an eyewitness report of this dramatic period of polymer science. Svedberg's decisive contribution to support of the macromolecular concept, as pointed out by Herman Mark, was in his measurements of molecular weight (mass), using the ultracentrifuge which he constructed in 1923 to 1926.

Svedberg's broad interests in research covered a wide field in physics, chemistry, biology and technology – and a life-long devotion to botany – as described by his close associate through several decades, Professor Sven Brohult, in his general lecture on "Svedberg as a Scientist".

The retrospective lectures on Svedberg's contributions to research present his early studies on colloid chemistry (P. Stenius), his pioneering and extensive protein research (K.O. Pedersen) and his studies on polysaccharides (I. Jullander) and synthetic polymers (P.O. Kinell). A personal account of the development of "The First Svedberg Centrifuges" in 1923-26 is given by one of Svedberg's students at this time, Dr. J.B. Nichols, who participated in this work both at the University of Wisconsin and at Uppsala University. One of the retrospective lectures, "Contributions of Svedberg to Nuclear and Radiation Research" (B. Larsson and S. Kullander) reviews Svedberg's life-long interest in this field, which resulted in the formation of the Gustaf Werner Institute as a department of Uppsala University.

The invited lectures on colloid and surface science, included in the Symposium Volume, are reviews of three active areas of research: surface-enhanced Raman scattering, which gives the intense colour of many colloids (M. Kerker), the colloidal behaviour of surfactant systems (B. Lindman and H. Wennerström), and the use of colloidal silica particles for centrifugal separation of biological materials (H. Pertoft and T.C. Laurent).

Ultracentrifugal sedimentation, pioneered by Svedberg for molecular weight (mass) measurements, is now a sophisticated technique supported by advanced theory (L.-O. Sundelöf) and developed as a source of more general information on macromolecules in solution (A.J. Staverman), especially for proteins in aqueous solution (H. Suzuki). Recent advances in thermal diffusion and sedimentation, another field of Svedberg's research interests, is also presented in the Symposium Volume (F.J. Bonner).

Macromolecular solutions and gels are extensively studied, e.g. by light scattering for molecular weight (mass) measurements (H. Benoit, M. Benmouna, C. Strazielle and C. Cesteros), and by Carbon-13 NMR for studies of association and phase structure in polymer gels (B. Lindstedt and P. Flodin). Remarkable success in analysis of proteins in solution is achieved by high performance electrophoresis (S. Hjertén and M. Zhu) and group-specific adsorbents (J. Porath, M. Belew, F. Maisano and B. Olin) as reported in the Symposium Volume.

Svedberg was a man of many interests and talents. As an introduction to an exhibition "The Svedberg - Scientist, Writer, Artist", opened on August 22 at the University Library "Carolina Rediviva", three short lectures were given: "Svedberg as an Author" by N. Gralén, "Svedberg as a Botanist" by C.-J. Clemedson, and "Svedberg as an Artist" by P. Ahlström.

They are included in the Symposium Volume to give perspectives on Svedberg´s remarkable achievements in many fields of human endeavour. This is further illustrated in the after-dinner speech on "Svedberg as a Promotor of Science and Technology" by N. Gralén, given at the Symposium Banquet at Uppsala Castle on August 23.

In the final chapter of the Symposium Volume, Svedberg speaks for himself on science, technology, society, and the human mind. It is in a lecture on "The Aims and the Means of Research" given by Svedberg on September 10, 1947, at the inauguration of an industrial research laboratory. This lecture presents Svedberg´s extensive knowledge of science and technology and his confidence in and enthusiasm for science as the basis for human progress. But he also expresses his deep concern and responsibility for the misuse of science applica-tions. The threat of nuclear devastation and the end of our civilization was clear to Svedberg in 1947.

The social programme of the Symposium programme offered a reception with a buffet supper on Tuesday, August 21, by invitation of Pharmacia AB, Uppsala, Sweden; a reception with supper courtesy of Uppsala Municipality and a concert featuring Swedish music by the Royal Academic Orchestra (Conductor, Director Musices, Dr. Carl Rune Larsson) in the University Main Building on Wednesday, August 22; a reception in the University Main Building to show the art and other treasures of Uppsala University by invitation of the President, "Rector Magnificus", Professor Martin H:son Holmdahl; and a buffet banquet in Uppsala Castle by invitation of LKB-Produkter AB, Bromma, Sweden, on Thursday, August 23. On Saturday, August 25, a sightseeing and excursion tour was made in Uppsala and its surroundings to the Cathedral, the old University Building ("Gustavianum"), the Bror Hjort Museum, to Old Uppsala with the Viking Kings´ Mounds and to Hammarby, the summer residence of the Swedish botanist Carl von Linné (Linnaeus), Professor at Uppsala University in "Medicine and Natural History" 1741-78.

ACKNOWLEDGEMENTS

The Svedberg Symposium was sponsored by the International Union of Pure and Applied Chemistry (IUPAC) as a Special Topics Symposium, by the Swedish Chemical Society, the Royal Swedish Academy of Sciences (KVA), the Royal Swedish Academy of Engineering Sciences (IVA) and by Uppsala University by placing its facilities at the disposal of the Symposium. The Symposium was made possible by generous grants from the Swedish Natural Science Research Council (NFR), the National Swedish Board for Technical Development (STU), the Carl Trygger Foundation for Scientific Research, the Research Foundation of the Swedish Cellulose Industry, and the Swedish Council for Planning and Coordination of Research (FRN).

As editor of the Symposium Volume, I would like to thank the authors for their con-tributions. The editing of the manuscripts has been done with a light hand and some repeti-tions seem inevitable. Svedberg´s work is seen from various angles by different coworkers and other scientists. The Symposium Volume is an account of Svedberg´s contributions and his importance in science. At present Professor Tore Frängsmyr, Uppsala University, is working on a biography of Svedberg. Svedberg´s personal recollections "Fragment" (356 pages) are deposited in the University Library of Uppsala University "Carolina Redivida" and have been made available to me by Mrs Margit Svedberg, which I gratefully acknowledge. Svedberg was modest, both as a person and about his own research. When he was awarded the Nobel Prize for Chemistry in 1926, he stated - as pointed out by Professor Martin H:son Holmdahl in his welcoming address to the Symposium - that the prize had been given too early. So far as is known, Svedberg is the only Nobel-Prize winner who has had this opinion. Our thanks are due to Professor Holmdahl for being an excellent host at Uppsala University and for contributing to the biography of The Svedberg.

Most of the manuscripts have been linguistically corrected by University Lecturer Donald MacQueen, Uppsala University, for which I am grateful.

July 1985 - The Royal Institute of Technology, Stockholm, Sweden

Bengt Rånby

Editor

I Introductory Lectures

1. Opening Address

B. Rånby, President of the Symposium

Department of Polymer Technology, The Royal Institute of Technology,
Stockholm, Sweden

The 100th anniversary of the birth of Professor The(odor) Svedberg takes us back to the turn of the century. He was born on August 30, 1884 at Fleräng Manor in the parish of Valbo close to the town of Gävle, about 100 km north of Uppsala.

Svedberg's childhood and school years were a happy time and also a time of great importance for his future development. He was the only child of Elias Svedberg, an industrial executive, trained as an engineer in mining and metallurgy, and his wife Augusta Alstermark. During Svedberg's school years, the family moved several times, from Valbo to Hommelviken in Norway where Elias Svedberg was head of a copper works and back to Sweden where Elias became head of various steel works in Västmanland, a mining and metallurgy district in the middle of Sweden. Elias Svedberg was a naturalist who collected minerals and plants and often took his son along on botanical excursions. This awakened an early interest in nature and science; "The Child is father of the Man" as William Wordsworth wrote in a poem. The five years 1895-1900 when Elias Svedberg was head of Karmansbo Steelworks were some of the happiest of Theodor's youth. He could make his own botanical excursions along the Hedströmmen River and also perform physics and chemistry experiments in his own little laboratory. I have seen plants in Svedberg's herbarium, collected and preserved at that time, 90 years ago.

Svedberg spent his school years in Köping Elementary School and Carolina Grammar School in Örebro (a gymnasium or high school). He had understanding and good teachers in both schools who let him carry on his own experiments in the physics and chemistry laboratories. In December 1903 Svedberg received his high school diploma (Baccalaureate) in a private examination in Gothenburg. He did not have time to wait for the regular examination period in Örebro half a year later.

As a university freshman at Uppsala University, in 1904, Svedberg started his research. He published his first research paper, on metal colloids, as early as 1905 at the age of 21. This meant the beginning of a new era in chemical research at Uppsala University, which until that time had been largely descriptive, preparative, and analytical. Svedberg had no major professor to teach and guide him. On his own, Svedberg developed new methods to prepare colloids in organic liquids using electrical discharge; he introduced physical measurements to study the size, size distribution, electrical charge and random motion of the colloidal particles; he applied mathematical theory to describe and interpret the results. In 1912, at the age of 28, Svedberg was made Professor of Physical Chemistry, the first professor of that subject in Sweden. This was a personal professorship for Svedberg.

Svedberg's research was immediately recognized as brilliant and given great attention by his colleagues, especially the chemistry professors Strömholm and Widman. This speaks well for the intellectual climate at Uppsala University at that time. A vivid and detailed account of these years is given in a paper by Professor J. Arvid Hedvall and published in the anniversary volume to Svedberg's 60th birthday. Hedvall was a contemporary fellow-student of Svedberg and his good friend through his lifetime.

Forty years later, in 1943, when I as a graduate student first met Svedberg, he was the dominant chemist and the most illustrious scientist in Sweden, widely recognized internationally by innumerable honorary degrees, awards and academy memberships. In his Institute, Svedberg was surrounded by a most distinguished, international group of scientists. Svedberg's students gradually obtained important positions in Sweden and abroad. My personal impression from these years is that of Svedberg's fertile mind, his enthusiasm and great interest in scientific problems, which hardly knew any boundaries between chemistry, physics, botany, and medicine. In addition, Svedberg was well informed in the technical, industrial, social, and cultural fields, he wrote books for the general public about energy, matter, and industry, he painted and he also read the current literature in French. Science and humanities, later described as the two cultures by C.P. Snow, were one culture for Svedberg. He was active and creative in both. Consequently, this IUPAC Symposium deals with Svedberg as a scientist in chemistry and physics, a botanist, a writer, an artist, and a photographer.

My lasting memory of Svedberg is of his brilliant mind and generosity with ideas. He took a personal interest in his students. You could discuss your research problems with him any time; he understood quickly, and he was always ready to help out by giving you his views and new ideas, and giving them generously. When you gave Svedberg a manuscript or a report of your research work, which may have taken you weeks to prepare, it seldom took him more than 10 to 15 minutes to read, evaluate and even correct it.   Such intellectual capacity is rare, and it fosters an exciting and creative atmosphere in a research laboratory.   I enjoyed it tremendously.

The character of Svedberg´s research is well described in a Chinese sentence which has been used to encourage young men and women to study science. It reads "Science is always young" and it was posted on large billboards in Canton (Guangzhou) during my visit there in 1979.   This oriental word of wisdom should be made a guiding principle for all scientific research.   Another sentence was also used in Canton at that time to attract students to science.   It read "Science is a flower forever".

These two sentences are particularly appropriate to quote at an anniversary symposium like the one we are now inaugurating.   The research work of Svedberg and his group was young science, and it is still flowering.   The ideas and results from Svedberg´s work have developed into new research and new applications, which we will hear more about during the next few days.

This symposium devotes the first half day to retrospective lectures.   The main part of the symposium (2 1/2 days) report new research in the Svedberg areas, i.e. hydrophobic colloids and surface science, hydrophilic and lyophilic macromolecules in solution and as gels, and nuclear chemistry related to biology and medicine.

I wish you all welcome to the Svedberg Symposium and declare the Symposium open!

REFERENCES

A. Tiselius and S. Claesson, The Svedberg and Fifty Years of Physical Chemistry in Sweden. Ann. Rev. Phys. Chem., 18, 1-8 (1967).
S. Claesson and K.O. Pedersen, The Svedberg 1884-1971. Elected for Mem. Royal Society 1944, Biograph. Memoirs of Fellows of The Royal Society, 18, 595-627 (1972). (Includes Svedberg´s complete bibliography.)
S. Brohult, Theodor Svedberg, 30.8 1984-26.2 1971.   Invald 9.4 1913.   Levnadsteckningar över Kungl. Vetenskapsakademins ledamöter, Nr. 184 (1982).
J.A. Hedvall, Från Nya Kemikums första tid.   The Svedberg 1884 30/8 1944, 681-725.   Almqvist & Wiksell, Uppsala (1944).

2. Polymers in the 1920's

Herman F. Mark

Polytechnic Institute of New York, 333 Jay Street, Brooklyn,
New York 11201 USA

NATURAL MATERIALS LONG EVADE RIGOROUS SCIENTIFIC ANALYSIS

Since the beginning of his existence, man has strongly relied on the use of natural organic polymers for food, clothing, and shelter. When he ate meat, bread, fruit, or vegetables and drank milk, he was feeding on proteins, starch, cellulose, and related polymeric materials; when he put on clothing made of fur, leather, wool, flax, and cotton, he used the same natural polymers; and when he protected himself against wind and weather in tents and huts, he constructed these primitive buildings of wood, bamboo, leaves, leather, and fabrics, which again all belong to the large family of organic polymers. In addition to the above-mentioned types, there are rubber, many resins, and bark.

Even later, when higher levels of civilization were reached, organic polymers were essential necessities in peace and war. All books in the famous library of ancient Alexandria consisted either of cellulose (paper) or protein (parchment), and they consist of these materials in all libraries of the world up to the present day. All transportation on land and sea throughout many centuries operated using wooden chariots and ships which were put in motion with the aid of ropes and sails made entirely of such celluloses as flax, hemp, or cotton. The music of all string instruments is produced by the vibrations of wooden, resin-treated boards; and all famous paintings together with many of the most valuable statues consist of cellulose, lignin, and polymerized terpenes in such materials as paper, canvas, wood, and paints. Bow and arrow are cellulose, lignin, resin, and proteins; catapults and siege towers were made of wood and moved with ropes and - until about 100 years ago - all sea battles were fought with wooden ships which were manoeuvered with the aid of cellulose sails and ropes.

While, in this way, natural organic polymers literally dominated the existence and welfare of all nations, virtually nothing was known about their composition and structure. In each sector - food, clothing, transportation, communication, housing, and art - highly sophisticated craftsmanship developed which was sparked by human intuition, creativity, zeal, and patience and led to accomplishments which will ever deserve the highest admiration of generations to come.

But even when the chemistry of organic compounds became a respectable scientific discipline in the early decades of the last century, the all-important helpers of mankind - proteins, celluloses, starch, and wood - were not in the mainstream of organic chemical research.

Why?

Because somehow they did not seem attractive at that time for a truly scientific study since they did not respond to the then existing methods for isolation, purification, and analysis. The experimental backbone of organic chemistry in those days was dissolution, fractional precipitation, and crystallization or distillation; it worked and still works with all ordinary organic compounds such as sugar, glycerol, fatty acids, alcohol, and gasoline but fails with cellulose, starch, wool, and silk. These materials cannot be crystallized from solution and cannot be distilled without decomposition.

This fundamental and embarrassing difference between the natural organic materials and the ordinary organic chemicals warned the chemists of the last century that there might be some essential and basic difference between these two classes of substances and that one would have to develop special, new, and improved experimental methods to force the second class into the realm of truly scientific studies.

NEW METHODS - NEW RESULTS - NEW CONCEPTS

In fact, during the second decade of this century there became available several new experimental methods which greatly accelerated the investigation of polymeric systems. Two of them permitted the study of such materials as cellulose, silk and rubber in the solid state and were essentially provided by the fundamental progress of physics at the turn of the century.

One of them was the scattering of X-rays by crystalline and semicrystalline systems which - according to M. von Laue and W.H. and W.L. Bragg - allowed determination of the positions of the atoms or ions in the investigated sample; it gave information on the geometry of its molecular structure in its native solid state.

The other method was the specific absorption of certain frequencies in the infrared region of the spectrum, which - according to quantum mechanics - revealed the existence of certain characteristic vibrations in the molecules and added dynamic information to the knowledge of geometry.

Two other, equally important methods implemented the solid state results by the study of polymeric systems in solution or suspension. The first was provided by the ultracentrifuge of The Svedberg and by the detailed description of its application and data evaluation in the classical book by Svedberg and Pedersen. This method permits the establishment of two average values of the weight of the solute, at the beginning essentially in aqueous systems, later also in organic solvents. The other method, initiated and substantially developed by Arne Tiselius was based on the motion of charged dissolved particles in an electric field and led to the technique of electrophoresis (and electro-dialysis) which also provided information on size and weight of solute particles in colloidal and polymeric systems.

During the same period refinements and improvements of osmotic cells made it possible to measure very small pressure differentials between solutions and the pure solvents. This led to number average molecular weights for starch in water and for rubber samples in benzene in the range between 15,000 and 25,000. Thus, at that time experimental techniques for the study of natural polymers in their native state and in solution became available and helped to open the door to a new world.

When I joined the Kaiser Wilhelm Institute for "Faserstoffchimie" in 1922, Professor R.O. Herzog, its Director, was engaged for two years with W. Jancke, M. Polanyi and K. Weissenberg in a systematic study of cellulosic fibres - ramie, flax, hemp and cotton. The chemistry of wood - cellulose, lignin and resins - was at that time dominated by Eric Haegglund and H. Erdtman in Sweden, K. Freudenberg and K. Hess in Germany and N. Haworth and J.C. Irvine in England. There existed an enormous material of observations and measure-ments, which, however, had not yet been condensed to a generally accepted concept of the two major components - cellulose and lignin. Two years earlier - in 1920 - Hermann Staudinger had published an article on "Polymerization" in which he claimed for several natural and synthetic polymers - rubber, polystyrene and polyoxymethylene - a chain structure in which the individual chain units were held together by normal chemical-covalent bonds. Staudinger did not mention cellulose in his paper but one year later - in 1921 - M. Polanyi on the basis of X-ray diagrams, and K. Freudenberg, as a result of kinetic measurement of cellulose degradation, indicated a long main-valence chain character of cellulose as a possible structure. Staudinger's insistence that the long chain concept termed as "macromolecular" theory was the only valid interpretation of the behaviour of natural and synthetic polymers, initiated one of the most animated and fruitful controversies in the chemistry of these substances.

LIVELY OBJECTION AND EVENTUAL CLARIFICATION

Objections to the macromolecular theory came essentially from three quarters.

The leading organic chemists of these days were essentially engaged either in studies of highly reactive natural substances such as alkaloids, vitamins and hormones or with the analysis and synthesis of important dyestuffs and pharmaceuticals; they were used to working with well-defined, crystallizable and soluble materials which could be readily purified and analysed. All these substances had molecular weights in the range of a few hundreds; solu-bility and fusability decreased rapidly with increasing molecular weight and it seemed to them incredible and unreasonable to assume that polystyrene, polyindene and rubber, all of which are readily soluble in benzene at room temperature, should have molecular weights of several hundred thousands. Certainly, this was only a subjective argument but it was backed up by such illustrious names as Richard Willstaetter, Paul Karrer, Heinrich Wieland, Max Bergmann and others.

Another group of scientists who disliked the macromolecular concept were the colloid chemists. Their position was that the assumption of such gigantic molecules was not necessary to explain the observed behaviour of viscosity, gel formation and solubility of such materials as polysaccharides, proteins and rubbers. All these phenomena may also be caused by agglomerates of small molecules which are held together not by normal chemical bonding but by association and complex formation. Evidently here was more than a subjective personal dislike, namely a realistic alternative explanation for the bulk of the known observations.

Even more dramatic was the position of a third group of scientists, namely the crystallographers. It had been established by X-ray analysis that the crystallographic elementary cells of cellulose (M. Polanyi), silk (R. Brill) and rubber (I.R. Katz), were so small that they could only accommodate a few monomeric units of these substances. Since supposedly - a molecule cannot be larger than the elementary cell - the hypothesis of macromolecular character was experimentally excluded for these materials.

In order to resolve this scientific controversy the German Chemical Society arranged a special symposium at the meeting of German Naturalists and Physicians in Duesseldorf.

On September 23, 1926, R. Willstaetter opened the famous meeting and called as first speaker, Professor Max Bergmann who presented several arguments for the aggregation structure of inulin and certain proteins. These arguments, together with a small elementary cell determination for silk were used by him to refute Staudinger's macromolecular postulate. Professor H. Pringsheim proceeded in a similar vein, referring to inulin and a few other polysaccharides. Both scientists quoted P. Karrer, K. Hess and R. Pummerer and referred to the well-known Werner complex compounds and to the high viscosity of many colloidal solutions. Staudinger presented extensive material on rubber and some synthetic polymers and based his contentions essentially on the high viscosity of polymer solutions and on some hydrogenation experiments with rubber, polystyrene, and polyindene. He had no data on cellulose and protein but suggested macromolecular structure for them, too.

The principal objection to Staudinger's proposal was the small elementary cell and that was just the topic about which I had to speak. I pointed out that there are cases, clearly discussed in several articles by A. Reis and K. Weissenberg, where the molecule can be larger than the elementary cell. This is always the case when true chemical main valences penetrate through the whole crystal. This could happen only in one direction, as in the case of cellulose and silk; in two directions, as in the case of graphite; and also in all three directions as in the diamond structure. Generally this means that small elementary cells do not exclude the existence of large molecules, but it also did not prove them either. It is a pity that neither the discussions nor the concluding remarks of Willstaetter have been published. As far as I can remember, Willstaetter thanked all lecturers and discussion speakers in friendly words and said, 'For me, as an organic chemist, the concept that a molecule can have a molecular weight of 100,000 is somewhat terrifying, but, on the basis of what we have heard today, it seems that I shall have to slowly adjust to this thought.'

In fact, soon there emerged more independent facts in favour of the macromolecular concept. O.L. Sponsler and H. Dore and K.H. Meyer and H. Mark put the compatibility of macromolecules with small crystallographic cells on a more quantitative footing not only for cellulose but also for silk, rubber and chitin. An extremely important contribution for the existence of macromolecules in solution came from The Svedberg through the systematic application of his ultracentrifuge first on proteins in aqueous systems and later together with R. Signer on polystyrene in organic solvents. Contribution was also made by W.H. Carothers in the laboratories of the Dupont Company through the controlled, stepwise synthesis of polyesters and polyamides.

Thus, by the end of the 1920's, Staudinger's concept was firmly accepted and a decade of intense research on cellulose, starch, proteins and rubber had wound up with the following fundamental results:

1. All investigated materials consist of <u>very large molecules</u>. Whereas the molecular weights of ordinary organic substances such as alcohol, soap, gasoline, or sugar range from about 50 to 500, the molecular weights of the natural organic building materials range from 50,000 to several millions, a fact which earned for them the name "giant molecules" or "macromolecules."
2. Most of them have the shape of <u>long flexible chains</u> which are formed by the multifold repetition of a base unit. One often refers to this unit as a "monomer" (monos is the Greek word for "one" and meros is the Greek word for "part") and to the macromolecules themselves as "polymer" (polys is the Greek word for "many").
3. If a sample consisting of regularly built, flexible chains is exposed to mechanical deformation, the individual macromolecules are <u>oriented</u> and show a tendency to form thin, elongated bundles of high internal regularity which are usually referred to as "crystalline domains" or, simply, as crystallites. Depending upon the nature of the materials and the severity of the treatment, a different percentage of the material undergoes crystallization whereas the rest remains in the "amorphous" or "disordered" state so that any given sample – fibre, film, rod, or disk – consists of two phases: amorphous and crystalline.

It was found that the crystalline domains contribute to strength, rigidity, high melting characteristics, and resistance against dissolution, whereas the amorphous areas impart softness, elasticity, absorptivity and permeability.

As soon as the study of natural polymers had started to establish these ground rules, chemists were strongly tempted to synthesize equivalent systems from simple, available, and inexpensive raw materials. The years from 1920 to 1940 brought ever-increasing successful efforts to:

1. Provide for more and cheaper basic building units – synthesis of <u>new monomers</u>.
2. Work out efficient equations to describe quantitatively the mechanism of <u>polymerization</u> and <u>polycondensation</u>.
3. Establish quantitatively the molecular weight and molecular microstructure to arrive at polymer <u>characterization</u>.
4. Explore the influence of the structural details on the different ultimate properties – <u>molecular engineering</u>.

Learning originally from nature and following up on the established principles, scientists and engineers succeeded in producing a wide variety of polymeric materials which outdo their original native examples in many ways and, in most cases, are much more accessible and less expensive. All this gave a tremendous lift to the important industries of man-made fibres, films, plastics, rubbers, coatings, and adhesives and made everybody's life richer, safer, and more comfortable. Statistics show that, at present, about 15 million tons of man-made fibres are produced and used in the world which amount to a total value of some 30 billion dollars. At the same time some 30 million tons of plastics - polyethylene, polypropylene, PVC, polystyrene, phenolics and many others - are made and used. They represent a total value of about the same order of magnitude. Similar figures hold for synthetic rubbers, coatings, and packaging materials. As a consequence, synthetic organic polymers have become a significant factor in the economy of all industrialized countries in the world.

3. Svedberg as a Scientist

Sven Brohult

Royal Swedish Academy of Engineering Sciences, Stockholm, Sweden

The(odor) Svedberg*) was enrolled at Uppsala University in January 1904. Before going to Uppsala he had been uncertain whether to study chemistry or biology, or more precisely, botany, which was his greatest interest. As it turned out, he chose chemistry, because he assumed that many problems in biology could probably be explained as chemical phenomena.

COLLOID CHEMISTRY

The Svedberg started his academic studies in the spring term of 1904. By September 1905 he had already finished his Bachelor of Science degree. He had decided to devote himself to the chemistry of colloids - probably influenced by Zsigmondy´s recently published book "Zur Erkenntnis der Kolloide" (Knowledge of Colloids). There is a distinction made between crystalloids and colloids. What distinguishes colloids, such as starch and proteins, is their inability to pass through parchment membranes (dialysis membranes), while crystalloids, such as sugar and salt, can pass through. Since one and the same substance, for example sulfur and metals, can exist both as a crystalloid and as a colloid, depending on the solvent, Svedberg finds it more correct to speak of crystalloidal and colloidal forms.

As early as December 1905 he published his first scientific paper, on the production of metal colloids in different media by electrical pulverization. Within this field there now followed a series of studies, which were summarized in his doctoral dissertation, in December 1907, ´Studien zur Lehre von den Kolloiden Lösungen´ (Studies of Colloidal Solutions). His doctoral work also included investigations of the movement of colloidal particles themselves, the so-called Brownian movement. It appeared plausible that the movements were a result of collisions of the fast-moving molecules of the liquid with the colloidal particles. Einstein had developed a theory about the phenomenon, which Svedberg tested by studying the movement of the particles in a Zsigmondy ultramicroscope. The results of Svedberg´s measurements confirmed the theory, and final proof had thereby been given of the existence of molecules. In 1912 Svedberg´s classic work "Die Existenz der Moleküle" (The Existence of Molecules) was published (ref.1).

In order to determine the formation and growth of particles in colloidal systems, it was found desirable to elaborate a method by which one could optically determine the concentration at different levels in a sedimenting system. The smallest particles that could be studied in the gravitational field had a diameter of two or three ten-thousandths of a millimetre (0.2 or 0.3 μm). To study smaller particles, the speed of fall had to be increased, and this could only be done with the help of centrifugal force. The ultra-centrifuge was knocking at the door.

AMERICA

World interest in colloid chemistry was great. At the University of Wisconsin in Madison there were plans for starting a Department of Colloid Chemistry. The Svedberg, regarded as one of the most distinguished colloid chemists of the time, was invited to come as a visiting professor to Madison, where he spent eight months in 1923. The Svedberg found many devoted disciples in America. Together with one of them, J. Burton Nichols, he constructed the first centrifuge in which the sedimentation of colloidal particles could be followed while the centrifuge was spinning. Another of his disciples, John W. Williams, later became Professor of Colloid Chemistry (physical chemistry) at the University of Wisconsin. Svedberg´s sojourn in Wisconsin proved to be of great importance to the development of colloid chemistry in the U.S.A.

During his visit to the U.S.A. Svedberg got many ideas for new work. It was here he realized that not only centrifuges were needed to characterize colloids, but methods such as electrophoresis and diffusion as well. The stay in Madison was a turning point in Svedberg´s research and consequently for physical chemistry in Uppsala.

---

*) During his early years at Uppsala University, Svedberg´s first name was shortened to "The" and pronounced as "Te" like in "Ten" (10).

THE ULTRACENTRIFUGE

Once home from America in late 1923, Svedberg immediately tackled the construction of the
ultracentrifuge (ref.2), which by definition could give sedimentation without convection.
Six months later the first measurements could get underway! In 1926 Svedberg received the
Nobel Prize for Chemistry for his work on "disperse systems" (i.e. colloid chemistry). In
his Nobel speech he described the first ultracentrifuge, and he stressed there the
importance this new instrument could have in colloid research especially. As early as 1925
such sensational results were attained with haemoglobin that Svedberg saw it as one of his
most urgent tasks to construct an ultracentrifuge well suited to probing the chemistry of
proteins. The principle of such a machine is simple: the earth´s field of gravity is
sufficient for a grain of sand, for example, to fall to the bottom of a beaker of water as
sediment. But for smaller particles and molecules to settle, it takes strong centrifugal
fields, which are created by the highspeed rotation of an ultracentrifuge. This construction entailed many difficult technical problems. In order to avoid explosion of the rotors
at the required speeds of 1,000–2,000 revolutions per second, the choice of steel was
important: it could be neither too soft nor too hard. Heat production could not be too
great; the rotor had to rotate in hydrogen under reduced pressure to decrease friction.
Tremendous demands were placed on the sector-shaped cell in which the sedimentation was to
take place. No convections were allowed to occur and no leakage of solutions. And the
entire construction had to allow observations to be made while the centrifuge was operating!
    It was to be a long and difficult road. It took more than fifteen years to arrive at
the perfect rotor. It turned out to be Rotor No. XXI, which was first used in 1939 – and
it´s still going strong! At a speed of 70,000 r.p.m., Rotor XXI yields a centrifugal force
that is 400,000 times stronger than gravity.
    Final success was no doubt partly due to the fact that Svedberg grasped the importance
of a precision workshop. To head this, he had found an extremely able precision mechanic,
Ivar Eriksson. Ivar Eriksson was a great asset when the going was rough. The and I once
came down after a rotor explosion. Ivar was already on the job, ´This is one hell of a
mess. I´ll need a week to clean up the debris. Then I´ll be ready for the next rotor´.

PROTEINS

The ultracentrifuge method was of tremendous importance to protein research. For about
fifteen years, 1925–1940, Svedberg and his collaborators devoted themselves to that research
(Svedberg´s protein period).
    Like most chemists, Svedberg had assumed that the lyophilic proteins were polydisperse
colloids like the lyophobic metal colloids he had studied earlier. At higher concentrations, proteins appeared to be waterswollen gels. Svedberg had planned to develop methods
for determining size-distribution of particles in protein sols. Imagine his surprise when
he found – using the ultracentrifuge – that the proteins he was observing were actually
monodisperse, that is, they had a well-defined molecular mass. Some of the proteins were
paucidisperse, that is, two or more definite molecular weight classes were simultaneously
present. By fractionating or changing the pH of the solvent, monodisperse solutions could
be obtained.
    Haemoglobin was the first protein to be studied thoroughly and whose monodisperseness
could be proved. The molecular weight was shown to be 68,000, which is four times greater
than the molecular weight of 17,000 derived from the iron content of haemoglobin. The
enzymes also turned out to be monodisperse.
    A great sensation occurred when the blue haemocyanin from the blood of the snail
Helix pomatia was centrifuged. Judging from the copper content of the protein, one would
have expected a molecular weight of 15,000 to 17,000. After just a few moments with the
haemocyanin solution in the ultracentrifuge, it turned out that the haemocyanin separated
out quickly, with a razor-sharp border surface. A rough calculation revealed that the
molecular weight had to be several million and that the molecules were of a uniform size.
It was the first time anybody had observed such uniform giant molecules. It was later shown
that this haemocyanin has a molecular weight of around ten million and is constant in a
broad pH interval.
    Svedberg and his collaborators studied a great number of proteins with molecular weight
from 34,000 (ovalbumin), 68,000 (haemoglobin), 70,000 (serumalbumin), 175,000 (serum-
globulin), up to millions (haemocyanin). Some haemocyanins yield association products
representing molecular weights of up to 80 million.
    Several proteins yield well-defined dissociation products when the pH factor is
changed. Thus the Helix pomatia haemocyanin produces half-, eighth-, and sixteenth-
molecules. This reaction is reversible: whole molecules are recreated if the solution is
returned to a neutral pH level. Haemocyanin consists of two kinds of molecules, of the same
weight, but with somewhat different properties; about one quarter of them can be split into
half-molecules by merely adding electrolytes. If both kinds of molecules are transformed
into eighth-molecules and are then returned to neutral pH, a mixture of molecules is
obtained in which the molecular weight is the same as the original molecule, but with
slightly different dissociation properties.

It has been found that haemocyanins from closely related species can produce mixed molecules. This is not the case, however, for species that are far away from each other in the biological chain. The door was open to a new science: molecular biology.

The Svedberg showed a great interest in haemocyanins. Partly as a joke, but also with a serious purpose, he founded the "Académie des Hémocyanines", with seventeen members. The Presidium of this Academy is now located in Louvain, Belgium, where there are a number of outstanding haemocyanin scientists. The President, René Lontie, has called a ceremonial special session of the Academy during this Symposium.

CARBOHYDRATES AND SYNTHETIC POLYMERS

The ultracentrifuge has also been of use in the study of other high molecular weight substances such as carbohydrates and synthetic polymers. It was extremely fortunate that interest was first focused on monodisperse proteins with their relatively round molecules rather than such substances as cellulose and starch, which were shown to be more complex; polydisperse and with long molecules, more or less extended chains in solution. The significant results with proteins no doubt spurred The Svedberg on in his search for a rotor which could withstand tremendous strain. Had he started with cellulose, which is polydisperse and has extended chains, the development of large-molecule chemistry would have been greatly delayed. Now that the ultracentrifuge had been developed for monodisperse substances, it was easier to tackle the study of polydisperse systems. It has also been possible to make significant contributions to the study of cellulose in its natural state, as well as to follow its breaking down in the pulping process of wood.

SVEDBERG AND RADIOACTIVITY

We have now followed Svedberg in his planned line of research from polydisperse colloids to the construction of the ultracentrifuge with its applications to monodisperse proteins and to natural and synthetic polymers. But there was another field of research, radioactivity, that had interested Svedberg ever since his days at the Carolina Grammar School in Örebro, where he was introduced to the work of Marie and Pierre Curie.

By studying the isomorphic condition of the radio-elements Ra, Th X, and Ac X, Daniel Strömholm, Professor of Inorganic Chemistry, and The Svedberg found in 1908-1909 that these three elements are chemically alike and should have the same position in the periodic system. Thus there are chemical elements with different atomic weights but with the same chemical properties. This happens to be the very definition of an isotope! Strömholm and Svedberg also put forward the hypothesis that the stable elements in the periodic system could be seen as mixtures of several elements – but not with identical atomic weights. Soddy, who received the Nobel Prize in 1922 "for his contribution to our knowledge of the chemistry of radioactive elements and his experiments regarding the prevalence and nature of isotopes", wrote roughly the following at the time. "It was not until I was preparing this Nobel speech that I found that Strömholm and Svedberg had explained the concept of the isotope in their work. But they hadn't given it a name!"

In the late 1930s Svedberg and his collaborators began to study the effect of different types of radiation on proteins. Haemocyanin was judged to be a suitable substance. I had the privilege of carrying out these experiments with Svedberg. We had hoped to be able to split the haemocyanin molecule at the pH where it was normally stable and then study the speed with which its parts were rejoined. The molecule was nearly split into well-defined halves and eighths. But we could not observe any re-uniting into whole molecules! Fission products obtained by irradiation and dissociation products obtained by a change in pH have the same molecular weight – but different chemical properties.

In 1945 Svedberg became involved in the establishment of a new institute of nuclear chemistry with a medical bias. In his commitment he was influenced by his former interest in radioactivity and his work on irradiation of proteins. In 1949 The Svedberg retired from his chair in physical chemistry and then continued until 1967 as head of the Gustaf Werner Institute for Nuclear Chemistry, a new department of Uppsala University. He devoted much of his time to the development of a synchrocyclotron and to the administration of the Institute. He also took an active part in the research – especially in the work in the fields of medicine and biology.

SVEDBERG AND INDUSTRY

In his book Research and Industry, published in 1918, Svedberg writes, "We must be vividly aware that research and industry are both equally necessary to us. And that both of these fields of endeavour must be developed in a planned manner. The four improductive and destructive years of the world war must be followed by years of increased production. Our scientists must not exclusively be kept busy with the retailing of knowledge, they must be given a chance to undertake fruitful production. We must get away from the idea that research is merely a way to while away leisure time." That appeal is just as urgent today – sixty years later!

Svedberg always had excellent relations with Swedish industry. In the 1930s substantial cooperation was established with the cellulose and the brewing industries, among others.

When Sweden could not get hold of the oil-resistant rubber needed by her armed forces during the Second World War, due to the blockade, Svedberg was asked by the State Commission for Industry in November 1941 to attempt to develop a process for the production of synthetic rubber in the shortest possible time by laboratory experiment. To be more specific, the task was to produce the type of synthetic rubber, produced from chloroprene by DuPont under the name of Neoprene. A small research facility was set up at the Institute, then a slightly larger facility in the industrial outskirts of Uppsala, and finally two factories, one at Ljungaverk and one at Stockvik were built (in the middle of Sweden, near Sundsvall). This Swedish oil-resistant rubber (named Svedoprene after Svedberg) proved to be of great importance for our defense. Among other applications, it was used for fuel lines and washers in our air force planes.

Many of you have no doubt had occasion to use - hopefully to your satisfaction - an LKB instrument. The creation of LKB Products Inc. can to a great extent be credited to The Svedberg; it is in all probability his greatest industrial accomplishment. The enterprise started, as a matter of fact, down in the basement of his Institute, with the legendary Ivar Eriksson, head of the workshop. For a short while LKB Products had a workshop in Uppsala down by Dragarbrunn Street under the supervision of Ivar Eriksson. When the company moved to Bromma, near Stockholm, Ivar was its adviser for the first few years. Svedberg himself played a very active part in the activities of LKB, both on its board of directors and its executive committee.

When LKB Products was founded in 1943, methods of separation analysis such as sedimentation, electrophoresis, and chromatography were the centre of attention at The Svedberg's Institute for Biochemistry under The Svedberg's disciple Arne Tiselius. And separation methods indeed became the successful core of production at LKB and still are today, even though other significant fields of enterprise have been added. And The Svedberg's advise has been followed: "Never produce an instrument, no matter how good it is, if it cannot be used for significant work in the service of science or technology".

LKB is now a well-established company with subsidiaries, Wallac in Åbo (Turku) in Finland and Biochrome in Cambridge (England) with sales nearing one billion (one thousand million) Swedish crowns. Its sales companies and agents are found all over the world. Last year LKB was introduced into the Swedish stock market.

SVEDBERG AS A BOTANIST

As was pointed out in the introduction, as a young student Svedberg found it hard to decide between chemistry and botany. Even though he opted for chemistry, he maintained a passionate interest in botany throughout his life. His botanical knowledge was enormous - easily the match of a professional botanist's - and it also found expression in a number of published articles from his pen. He possessed a rare eye for everything involving flora and on top of that the ability to tell at a glance what a certain locale would have to offer a botanist. He was seldom wrong. It was his ambition to see all of the main species of Phanerogamia and Vascular Cryptogamia in Sweden. And he succeeded, but it took nearly his entire life to do so. The Svedberg made significant contributions concerning the migration of plants into Scandinavia.

In his late years, The had the opportunity to make two botanical dream journeys, which he had long yearned to make, to Greenland and to Spitsbergen. Both of these trips are described in the botanical literature.

SVEDBERG AS A CULTURAL FIGURE

The Svedberg was a man of many cultural interests. He was a great admirer of Strindberg, and in Forum (1918) he writes, "In my youthful enthusiasm and in exaggerated admiration of Strindberg, I had undertaken to go through his chemistry and thought I would recreate some of his experiments." The Svedberg's rationale was that a person of such originality as Strindberg ought to have arrived at something of value among all the dross in his chemical writings. In 1908 and 1909 Svedberg corresponded with Strindberg, checked his chemical writings and repeated some of his experiments. When Strindberg noticed that the results were not confirming his assumptions, he severed their connections. When Svedberg visited Strindberg at the "Blue Tower" on Queen's Street in Stockholm on another errand, he proposed that Svedberg be allowed to edit and publish his unprinted chemical manuscripts. He saw the reluctance in Svedberg's eyes, and the matter was never discussed again.

The reviews of Strindberg's writings after his death in 1912 contained admiring assumptions about his scientific work. Svedberg felt that he ought to go along with requests that he make public his correspondence and contacts with Strindberg. After reviewing Strindberg's chemistry anew, he published the 1918 article in Forum "Strindberg as a Chemist". I quote Svedberg's final verdict: "An analysis of Strindberg's chemical writings yields, apart from the illumination of details, a wealth of errors and the non-existence of new ideas, another, more general, illumination. The sum of all the details is the conclusion: Strindberg was no scientist. In his innermost soul he was an anathema to all true research. Therefore his toil and striving along these lines were in vain, chasing windmills. The value of Strindberg´s chemical writings is entirely of a literary and psychological nature. In particular, they ought eventually to be of extraordinary interest in determining his spiritual development. His scientific studies have without question served to fertilize his writing, given it direction in the form of fresh, new images and in a rather unusual way brought him into intimate contact with the world around him."

But Svedberg´s interest in literature was wide. The Svedberg´s library was well equipped with an exquisite collection of modern and ancient Swedish literature, often original editions. He was one of our greatest connoisseurs of French literature, both prose and poetry. He greatly enjoyed reading French poetry in the original language and was always eager to discuss the latest works. In his foreign travels, he often took the opportunity to visit museums and art exhibitions. The Svedberg was himself a painter, especially of landscapes. His paintings are admired and are now seen as rarities. The development of atomic energy and biology in the 1940s and 1950s inspired him to design cloth for draperies, named Atomics and Genetics.

I have had the privilege of going through some of The Svedberg´s remaining notes, "Fragments". It is hardly surprising, considering Svedberg´s great interest in poetry, to find there poems, above all from 1917 and 1918. The mood of several of them was no doubt influenced by the war, but there are also several in which The expresses the memories of his happy childhood years in the Bergslagen mining district, for example, "Autumn Mood from Karmansbo" and "September":

September

The Manor garden has just been harrowed by
The first frost. The dahlia´s heavy
Buds hang slack and the tomato´s
Leaves are blackened. The aster stands there
Plucky and chilly and against a high, blue
Sparkling September sky,
The hemp raises its tough
Green branches. The big garden path
Shines white and red with
Fallen fruit. The garden boy comes along,

Casually squinting in the morning sun
With a basket to pick
Clear apples under
The trees. The fog from the river
Is still drifting in layer after layer of
Light veils through the park.
The arbor´s weeping ash slowly loses
Its leaves. One after another they sink
Down in the fog and are gone.

(Svedberg the botanist, the author and the artist has been dealt with by others during the Symposium.)

Few Swedes have meant so much to the development of chemistry as The Svedberg. He had innumerable disciples, both abroad and in Sweden. It is a long list, with many well-known names. The time allotted me does not permit an enumeration of them all. It is a source of great satisfaction for the organizers of this Symposium to have the honour of welcoming so many of The´s disciples from far and near. It is a special pleasure for us to see two of his collaborators from his time in America - J. Burton Nichols and Harold P. Lundgren.

The Svedberg had a divine gift for enthusing his disciples. If finding the answer to an important problem was at stake, there was no difference between workday and holiday, nor day and night.

One evening (or rather night) I was centrifuging haemocyanin from the Helix pomatia. Time and time again The came down to the ultracentrifuge, curious and impatient. The result of the experiment turned out to be different from what we had expected. But The came rushing down and exclaimed, "It´s even more remarkable than we thought. My dear Sven, is there anything on this earth more beautiful than a haemocyanin molecule?"

REFERENCES

T. Svedberg, Die Existenz der Moleküle, Leipzig (1912).
T. Svedberg, Prix Nobel 1926, Stockholm (1927).
T. Svedberg and K.O. Pedersen: The Ultracentrifuge. Clarendon Press, Oxford (1940).
The Svedberg 1884-1944 (Anniversary Volume). Almquist and Wiksell, Uppsala (1944).
S. Claesson and K.O. Pedersen, Biographical Memories of Fellows of the Royal Society (with complete biography), Vol. 18 (1972).
S. Brohult, Theodor Svedberg: Minnesteckning vid Kungl. Vetenskapsakademiens högtidssammankomst den 31 mars 1984.

II Retrospective Lectures on Svedberg's Contributions

4. Svedberg's Early Studies in Colloid Chemistry

Per Stenius

Institute for Surface Chemistry, Box 5607, 114 86 Stockholm, Sweden

Abstract - A review is given of The Svedberg's early work in colloid chemistry, roughly covering the years 1905-1920. During these years, Svedberg made extensive studies of methods of preparing colloids, Brownian movement, sedimentation, diffusion and light adsorption. The monographs "Methoden zur Herstellung Kolloider Lösungen Anorganischer Stoffe" (1909) and "Die Existenz der Moleküle" (1912) summarize much of the work that is still of importance. Svedberg's interpretation of his studies of Brownian motion is discussed.

To most of us, at least to those of us belonging to the post-war generation, The Svedberg's name is connected with the ultracentrifuge and the first quantitative determinations of the molecular weights of macromolecules. Indeed, I believe that anyone can verify by a simple poll of opinion that most colloid scientists (not to mention chemists in general) believe that Svedberg received his Nobel Prize in recognition of the significance of the results obtained with the ultracentrifuge. One is easily strengthened in this belief by Svedberg's own Nobel lecture (refs.1, 2) which is concerned with the exciting new findings using this instrument.

Undoubtedly, the really lasting impact of Svedberg in colloid science now seems to be connected with the development of ultracentrifugal, diffusional and electrophoretic methods of characterizing colloids. These topics are amply dealt with by other authors in this volume.

Yet, the Nobel Prize in Chemistry which Svedberg received in 1926, was given at a time when even the first studies of proteins with the ultracentrifuge had been only briefly published. As is clear from the endowment speech given by Professor Söderbaum, the secretary of the Royal Swedish Academy of Sciences, Svedberg's prize was given in recognition of his work on disperse systems and, in particular, for the confirmation of Einstein's theory of the Brownian movement which was one result of this work. Indeed, the Nobel prizes in 1926 constitute a broad recognition of the importance of the work of colloid chemists. The endowment of the 1925 prize in chemistry had been postponed and it was now given to Richard Zsigmondy while Jean Perrin and The Svedberg received the prizes for 1926 in physics and chemistry, respectively, in recognition of work that preceded the ultracentrifuge by almost two decades. This massive acknowledgement of the achievements by these three scientists makes it clear that the importance of such studies in the first decade of this century can hardly be overemphasized. Indeed, they were able definitely to settle the long-lasting discussion concerning the kinetic theory of heat and the existence of molecules as kinetic units.

Einstein published his theory of Brownian motion in 1905 (ref.3). The theory is based on the assumption that Brownian motion is caused by the molecular motion of heat leading to random collisions between the unobservable molecules and the microscopically observable colloidal particles. His well known result is that a microscopically visible body suspended in a liquid of viscosity η at temperature T will undergo random motion in such a way that the root-mean-square displacement $< \underline{x}^2 >$ in the direction $\underline{x}$ in the time-interval $\underline{t}$ will be given by

$$< \underline{x}^2 >^{\frac{1}{2}} = (2\underline{Dt})^{\frac{1}{2}} \qquad (1)$$

where for a spherical particle of radius $\underline{a}$ the diffusion coefficient $\underline{D}$ is given by

$$\underline{D} = \frac{kT}{6\pi\eta\underline{a}} \qquad . \qquad (2)$$

($\underline{k}$ = the Boltzmann constant)

Einstein himself pointed out that if his theory could be experimentally verified, this would constitute a decisive proof in favour of the kinetic-molecular conception of heat.

Without any knowledge of Einstein's work, Svedberg had started studies of the Brownian motion of colloidal metal particles in 1904-05. On reading Einstein's article, Svedberg immediately understood the relevance of his own experiments with regard to this theory.

Thus, Svedberg in 1905 claimed (refs.4, 5) and later persistently maintained (ref.6) that
his studies of Brownian motion confirmed eqn (1) and hence constituted proof of the
existence of molecules. This was, indeed, a result worthy of recognition. And, although
Svedberg´s interpretation of his experiments later was criticized, it seems that it actually
was recognized at that time.

A couple of years later, Jean Perrin and his co-workers (ref.7) published studies of
the temperature-dependence of the mean displacement of cinnabar particles due to Brownian
motion that very clearly demonstrated the validity of Einstein´s theory. Nowadays, in
colloid chemistry textbooks, the work of Perrin is exclusively cited as having yielded the
definite confirmation of Einstein´s theory.

In 1904 The Svedberg matriculated at the University of Uppsala at the age of 19.
Already in his school years, however, he had been intensely interested in chemical preparations and he passed the courses and examination for the Fil Kand (BSc) degree in less than
two years. Despite the intense studies for his degree, he also found time to peruse more
advanced literature. His interest in colloids was aroused by Nernst´s Theoretische Chemie
which he reportedly read during his "leisure hours". Zsigmondy´s book on colloid chemistry
was probably also an important source of inspiration (refs.10, 11). Svedberg felt convinced
that the study of colloidal systems would have great impact on the understanding of the
processes in living matter and therefore decided to devote his first scientific work to the
preparation of well-defined colloids. This is also the topic of his first paper published
in 1905 (ref.8).

Svedberg obviously from the very beginning clearly realized the importance of
quantitative measurements on reproducibly prepared colloidal model systems and the
scrupulous experimental attention to detail required to obtain such systems; a problem that
unfortunately to this day is not always sufficiently appreciated by many colloid scientists.
In his first and several subsequent papers he describes the preparation of metallic
organosols. In Bredig´s method of preparing colloidal metal dispersions, an electric arc is
passed through a mixture of solvent and powdered metal. Svedberg modified this method by
using alternating rather than direct current and careful investigation of the optimal
conditions for the production of stable colloids. One of his photographs of the arc is
shown in Fig. 1.

These modifications made possible the preparation of a large number of new colloids and
already in his thesis (ref.5) Svedberg described the preparation of sols of more than 30
metals in organic solvents, mainly isobutanol and ethyl ether. The synthetic work was
expanded in the following five to ten years, and in 1909 a large and detailed monograph on
the preparation of colloids was published (ref.9). This monograph has become classical and
still is a very useful reference work on the preparation of metal sols by electrical
discharge or chemical reaction.

Fig. 1. Photograph of the preparation of a dispersion of cadmium in
ethyl ether by the use of an oscillatory discharge. (The Svedberg,
Nova Acta R. Soc. Sci. Uppsala, (4) **2**, 1-160 (1907).)

Having obtained suitable systems for quantitative investigation, Svedberg then proceeded to study the stability (mainly the conditions for obtaining stable sols) and Brownian motion of the colloidal particles. In the latter studies the ultramicroscope recently developed by Zsigmondy and Siedentopf played a central role. The original experimental setup, shown in Fig. 2, was not very complicated.

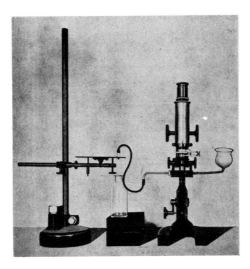

Fig. 2. Svedberg's original experimental setup for the study of colloids in the ultramicroscope. (The Svedberg, Nova Acta R. Soc. Sci. Uppsala, (4) **2**, 127 (1907).)

As Fig. 2 shows, an important detail of this apparatus is that the dispersion under observation slowly flows through the observation cell during the experiment. The reason for this arrangement was the conception that in the Brownian motion the particles moved back and forth between extreme positions. Svedberg's ingenious idea was that such a motion in a flowing liquid observed by a static observer would be seen as more or less sinusoidal, and hence the time required for the particles to move between the two extreme positions could be determined from the wavelength of the observed sinusoidal motion and the flow rate of the liquid.

The source of the idea that the Brownian motion was "oscillatory" can be found in Zsigmondy's description of the phenomenon in his 1904 book (ref.10).

On the basis of experiments carried out with platinum sols, Svedberg claimed that the observed trajectories were indeed sinusoidal with constant wavelength and mean amplitude.

He found that the amplitude of the oscillations was inversely proportional to the viscosity of the dispersion medium. Very briefly, Svedberg then claimed that this result confirms Einstein's theory in the following way. The particle moves back and forth between the extreme positions of the observed sinusoidal curve in one period of oscillation (τ); the distance covered is four times the amplitude of the oscillation ($\underline{A}$). Substitution of $4\underline{A}$ for $\langle \underline{x}^2 \rangle^{\frac{1}{2}}$ and τ for $\underline{t}$ in eqn (2) gives

$$4\underline{A} = \tau^{\frac{1}{2}} \left(\frac{kT}{3\pi\eta\underline{a}} \right)^{\frac{1}{2}} \tag{3}$$

Svedberg's careful preparation of colloids allowed him to assume that the particle size was constant and hence

$$\underline{A} = \underline{c}_1 (\tau/\eta)^{\frac{1}{2}} \tag{4}$$

or

$$\underline{A}/\tau = \underline{c}_2/\underline{A}\eta . \tag{5}$$

His experimental results showed that the ratio $\underline{A}/\tau$ which Svedberg called the absolute velocity of the particles, was constant, and hence, in accordance with his results,

$$\underline{A}\eta = \text{constant.} \tag{6}$$

A more detailed discussion of Svedberg's interpretation has been given by Kerker (ref.12).

Svedberg's arguments came under immediate criticism. First of all, Einstein himself (ref.13) pointed out that due to the random nature of the Brownian motion it is not possible to determine an absolute velocity of the particles. Later, Perrin (ref.14) stated (as is indeed the case) that no oscillatory movement of the particles can actually be observed. It should be stressed, that it was from the beginning quite clear to Svedberg that the movements were not oscillatory in the sense that there would be a restoring force proportional to their displacement acting on them (refs.14, 16). He also definitely established that Brownian motion was not due to electrical charge on the particles. Svedberg, however, continued to claim that the observed motions in the moving liquid could be seen as sinusoidal.

From 1909 onwards, he and his co-workers made a series of renewed experiments with the motion of the particles recorded by a moving camera and both in his book of 1912 on the existence of molecules (ref.6) and much later in his lectures on colloidal science at the University of Wisconsin in 1923 (published as a book in 1924 (ref.17) he still cited his proof of Einstein's equation. Although this proof, naturally, no longer occurs in colloidal chemistry textbooks, it did have considerable impact on contemporary scientists. Thus, the probably most well known unbeliever in molecules as kinetic units, Wilhelm Ostwald, apparently actually did change his mind as a consequence of Svedberg's results. In his review of Svedberg's thesis (ref.15), Ostwald states, that he has finally found experimental evidence for the kinetic theory that so many adherents of the atomic theory have for so long failed to produce. Such a recognition of the work of a 23 year old scientist by one of the founding fathers of physical chemistry must have had a great impact indeed.

During the decade following the studies of Brownian motion and preparation of colloids Svedberg, having obtained a personal chair in physical chemistry at Uppsala University in 1913 (the first chair in physical chemistry in Sweden), devoted the major part of his research work in colloid chemistry to the study of the physico-chemical properties of colloidal systems, in particular the size distribution of the particles and the concept that as the particle size grows there is a continuous transition from "crystalloid" (i.e. molecularly dispersed) solutions to colloids. In 1911 (ref.18) Svedberg had already pointed out that measurements of Brownian motion and sedimentation in the ultramicroscope could in principle (although only with extremely time-consuming procedures in practice) be used to determine size distributions. Methods that were also tried were fractional coagulation, weighing of sediments and sedimentation in the gravitional field. Above all, during these years, he inspired young scientists (for example, the classical work of Odén on sulfur sols and the sedimentation balance (ref.19) and Westgren's studies of Brownian motion (ref.20), to further work in colloid chemistry and established his fame as one of the most brilliant experimentalists in the world in the field of colloid science.

In his studies of sedimentation, it became clear to Svedberg that the variation of the concentration with height, which is related to the size distribution, must be recorded by optical means. The optical properties of colloidal systems were extensively used not only for this purpose but also to demonstrate the transition from crystalloid to colloid. Gradually, Svedberg realized that in addition it was necessary to enhance sedimentation by using a centrifugal field. The invitation by Professor J.H. Matthews to lecture and organize research in colloid science at the University of Wisconsin in 1923 gave Svedberg the opportunity to begin realizing his ideas and led to the work for which he is now best remembered.

It seems to me very natural that Svedberg, when awarded the Nobel Prize in 1926, chose to devote his lecture to the ultracentrifuge and did not go into his earlier work. His brilliant experimental imagination shines through very clearly, however, already in his first papers. It cannot be denied that Svedberg's early studies in colloid chemistry yielded practical knowledge about colloidal systems of lasting importance. His continuing effort to make quantitative studies of these systems created a foundation without which the grandiose achievements that followed the construction of the ultracentrifuge would not have been possible.

REFERENCES

1   T. Svedberg, Nobelföredrag hållet i Stockholm den 19 maj 1927 Prix Nobel 1926, Stockholm, 1-16 (1927).
2   T. Svedberg, Nobelvortrag gehalten zu Stockholm am 19. Mai 1927. Kolloid-chem. Beih., 26, 230-244 (1927).
3   A. Einstein, Ann. Phys., 17, 549-560 (1905).
4   T. Svedberg, Z Elektrochem., 12, 853-860 (1906); ibid. 12, 909-910 (1906).
5   T. Svedberg, Zur Kenntnis der Stabilität kolloider Lösungen, Nova Acta Roy. Soc. Sci. Uppsala, (4) 2, 1-160 (1907).
6   T. Svedberg, Die Existenz der Moleküle, Akademische Verlag, Leipzig, (1912).
7   J. Perrin, Ann. Chim. Phys., 18, 1 (1909).
8   T. Svedberg, Ber., 38, 3616 (1905).
9   T. Svedberg, Die Methoden zur Herstellung kolloider Lösungen anorganischer Stoffe, Theodor Steinkopf, Dresden (1909), new editions 1920 and 1922.
10 R. Zsigmondy, Zur Kenntnis der Kolloide, Jena (1905).
11 S. Claesson and K. Pedersen, Biographical Memoirs of Fellows of the Royal Society, 18, 595 (1972).

12 M. Kerker, Isis, **67**, 190 (1976).

13 A. Einstein, Z. Elektrochem., **13**, 41 (1907).

14 J. Perrin, Ann. Chim. Phys., **18**, 1 (1909).

15 W. Ostwald (W.O.), Z. Phys. Chem., **64**, 508 (1908).

16 T. Svedberg, Arkiv Kemi, **B2**, 34 (1906).

17 T. Svedberg, Colloid Chemistry, ACS Monograph **16**, Chemical Catalog Co, New York (1924).

18 T. Svedberg and K. Estrup, Z. Chem. Ind. Kolloide, **9**, 259 (1911).

19 S. Odén, Der kolloide Schwefel, Nova Acta Roy. Soc. Sci. Uppsala, (4) **3**, 1 (1913).

20 A. Westgren, Untersuchungen über die Brownsche Bewegung besonders als Mittel zur Bestimmung der Avogadroschen Konstante, Almqvist & Wiksell, Uppsala and Stockholm (1915).

5. Svedberg and the Proteins

Kai O. Pedersen

Institute of Physical Chemistry, Uppsala University, Uppsala, Sweden

Abstract - After Svedberg´s return to Uppsala from Madison in 1924, he
built an electrically driven centrifuge, where he could optically follow
convection-free sedimentation in a sector-shaped centrifuge cell. When
a haemoglobin solution was run in this centrifuge, it sedimented and
after equilibrium was reached, a molecular weight (mass) of 68,000 could
be calculated; it was constant all the way through the cell. Svedberg
was excited! This was the first time a monodisperse protein had been
observed. Could it be that the proteins had well defined monodisperse
molecules? He came to the conclusion that the best way to study this
would be to study the sedimentation curves in strong centrifugal fields
and calculate the diffusion coefficient from the blurring of the
sedimentation curve. It should be the same as in free diffusion experiments. If sedimentation velocity experiments should be used, it would
be necessary to apply a much more intensive centrifugal field such as
70,000 to 100,000 g or a 15 to 20 times stronger field than used in the
first equilibrium experiments. Before Svedberg could construct such a
new centrifuge, a number of new problems, such as technique, safety, and
financing, had to be solved.

Svedberg´s first paper was published as early as December 1905. It dealt with the preparation of organo-sols of various metals. To study these metal sols he built a Zsigmondy-
Siedentopf ultramicroscope together with Carl Benedicks, at that time docent in physical
chemistry. Two years later Svedberg had completed his dissertation: "Studien zur Lehre von
den Kolloiden Lösungen". It was presented in December 1907, and he became docent in
physical chemistry at Uppsala University. He could now have his own students. Together
with his pupils Svedberg started a comprehensive study of the properties of colloidal
solutions. Svedberg was also interested in the formation and growth of the particles in
metal sols. Together with Herman Rinde he developed an optical method by which he could
follow the variation in concentration by height in a sedimenting column, and from the
variation in concentration they could calculate particle sizes and particle size distribution at different heights in the column from the rate of movement of the different levels in
the sedimenting system.
Svedberg was very productive in these years, and by the time he got a royal appointment
as Professor of Physical Chemistry at Uppsala University in 1912, he had already published
more than forty papers.
The smallest particle Svedberg and Rinde could study by their method had a diameter of
200 nm. For particles smaller than that the rate of sedimentation was too small to be
measured with accuracy. If smaller particles were to be measured, it would be necessary to
make use of centrifugal fields. A first draft for such a method was made in July 1922.
Here Svedberg combined sedimentation in a centrifugal field with ultramicroscopic
observations.
At this time interest in colloid science was very prominent, and at the University of
Wisconsin at Madison it was planned to start a division for colloid science. It was therefore quite natural to approach one of the leading colloid scientists in Europe. Svedberg
was thus invited to come as a guest professor to the University of Wisconsin. Svedberg was
excited, and during the autumn of 1923 he made preparations for a research programme to be
carried out at Madison. It included centrifuge experiments, electrophoresis, diffusion, and
electrical colloid synthesis.
In Madison he found some very enthusiastic pupils. Together with one of them, Burton
Nichols, he constructed the first centrifuge where the sedimentation of the particles could
be followed either by direct optical observation or by photographs taken during centrifugation. However, since the sedimentation took place in ordinary cylindrical centrifuge tubes,
the particles were carried down not only by sedimentation but also by convection along the
wall of the tubes; no exact values for sedimentation velocity of the particles could be
determined.
The seven months in Madison were extremely stimulating for Svedberg, and he returned to
Uppsala full of new ideas. But he was not only thinking about centrifuges. He was also
considering electrophoresis and diffusion together with other problems of colloid science.

The time in Madison meant a turning point in Svedberg's research and in physical chemistry in Uppsala.

The most urgent problem now was to construct a special ultracentrifuge. On the Atlantic in August 1923, on the way back to Uppsala, Svedberg made sketches of rotors and centrifuges which he hoped would result in stronger centrifugal fields than those of the Madison centrifuge. They should have sector-shaped cells, and the centrifuge should run without vibration.

In January 1924, Svedberg and Rinde started to build the first ultracentrifuge. Sector-shaped cells were cut into a 10 mm thick round glass plate (diameter 12.4 cm). On each side 5 mm thick glass plates were fastened with the Kotinsky cement. These glass plates were placed at the top of a heavy brass rotor, which rested on the top of the spindle of a rebuilt milk separator. The rotor was surrounded by a gas-tight container in which it could rotate in a controlled atmosphere. At the beginning, there were quite a number of problems with convections and vibrations. However, after the rotor was allowed to spin in a reduced hydrogen atmosphere and after some changes in the driving system were made, Svedberg and Rinde could start experiments with gold colloids. In July 1924 the first ultra-centrifuge manuscript was posted to the editor of the Journal of the American Chemical Society. The authors emphasized the importance of this new instrument.

Like other colloid chemists and most other chemists, Svedberg assumed that proteins were polydisperse lyophilic colloids with no definite molecular weight. He had planned to work out methods for determining the distribution of the particle sizes in protein sols. He had tried to centrifuge egg albumin without any result. Then in September 1924 Robin Fåhraeus came to Uppsala to study with Svedberg. Fåhraeus was docent in pathology at the Caroline Institute in Stockholm. They started to centrifuge active casein from milk and found that it contained a broad spectrum of particle sizes from 10 nm up to about 70 nm. This was just what Svedberg had expected!

After the experiments with casein Fåhraeus proposed that they should try with haemo-globin. Svedberg was extremely sceptical! The egg albumin had not sedimented when he had tried; and according to S.P.L. Sørensen's osmotical measurements at the Carlsberg Laboratory in Copenhagen, it was supposed to have an average molecular weight of 34,000, while on the other hand haemoglobin should have an average molecular weight of about 17,000 according to its iron content. However, Fåhraeus thought it was well worth seeing if haemoglobin would sediment. In any case, on October 16, 1924, they started an experiment. After some hours Svedberg went home and let Fåhraeus watch the centrifuge.

At 2 o'clock in the morning Svedberg was awakened by a telephone call from Fåhraeus, who told him that he saw a dawn in the cell! Svedberg rushed back to the laboratory. It was quite evident that the colour was much lighter at the top of the cell: the haemoglobin had started to sediment. Some hours later a crack developed in the cell window, and the haemoglobin solution leaked out!

A new and better cell construction was made, and on November 12-14, 1924, the first successful sedimentation equilibrium experiment with a protein was carried out.

It showed that the molecular weight of haemoglobin was about 68,000 and the molecular weight was the same from the top to the bottom of the cell. This was the first time that the monodispersity of a protein had been demonstrated. This was a great surprise not only to Svedberg but also to many others. The experiment started a new epoch in protein chemistry.

The experiments were continued, and the first paper about the molecular weight of haemoglobin was sent to the Journal of the American Chemical Society in July 1925 (ref.1). The authors wrote: "The lack of a reliable method for determination of the molecular weight of substances possessing a very complicated structure has been a serious obstacle in the progress of our knowledge of the chemistry of the proteins. In the present paper such a method will be proposed, and its use will be illustrated by a few preliminary measurements on haemoglobin".

Svedberg wondered if it could be so that proteins had a well-defined molecular weight and consisted of real molecules, not of micelles having different sizes, as generally assumed by chemists at that time.

After the sensational result with haemoglobin it became an urgent matter for Svedberg to study proteins. He considered it important to change the experimental method from sedimentation equilibrium to sedimentation velocity. A comparison of the diffusion coefficient calculated from the sedimentation boundary between the protein solution and the buffer in the centrifuge cell and from the boundary in free diffusion experiments would immediately show if the substance under investigation was polydisperse or had equal-sized particles.

If sedimentation velocity experiments were to be used, it would be necessary to use a much more intense centrifugal field, such as 70,000 to 100,000 g or a 15 to 20 times stronger field than was used in the first equilibrium experiments.

Before Svedberg could construct such a new centrifuge, a number of new problems, such as, technique, safety and financing had to be solved.

Svedberg had earlier contacted Mr Fredrik Ljungström at the Ljungström Steam Turbine Co. in Stockholm. He had recommended the use of oil turbines to drive the centrifuge rotor as this would simplify the lubrication problems. He also contributed other ideas and had one of his younger engineers, Alf Lysholm, assist Svedberg in the construction and testing of the new ultracentrifuge, which was built during the spring and summer of 1925 at the workshops of Ljungström Steam Turbine in Stockholm. The financial problem was solved by a research grant of 25,000 Swedish crowns from a new foundation for medical research, "Therese and Johan Anderson´s Memorial".

The installation of the ultracentrifuge went on the whole autumn of 1925, and not before January 10, 1926, could the first test run be started. However, it was a disappointment: instead of the expected 40,000 r.p.m. not more than 19,000 r.p.m. was attained.

A large number of unexpected technical problems had to be solved. The turbines and the bearings had to be reconstructed, the oil system was rebuilt. The gas friction around the rotating rotor was much reduced by having the rotor spin in an atmosphere of hydrogen at low pressure. Finally on April 7, 1926, the ultracentrifuge was run at 40,100 r.p.m. (ref. 2).

Still a number of small problems remained to be solved. They were gradually ironed out, however, and Svedberg could concentrate on the problem that interested him more than anything else: were proteins monodisperse, well-defined substances? Burton Nichols came to Uppsala from Madison, and a number of experiments were run on CO-haemoglobin. They all showed uniformity. Svedberg was excited.

Just at this time the Royal Swedish Academy of Science announced on November 12, 1926, that the year´s Nobel Prize for Chemistry had been conferred on The Svedberg for his work on disperse systems. At the same time two other colloid scientists were awarded Nobel Prizes: Richard Zsigmondy received the 1925 Prize for Chemistry and Jean Perrin the 1926 Prize for Physics.

The Nobel Prize was of course a very great stimulus for Svedberg, and it became much easier for him to get research grants, so he could start planning further development of his ultracentrifuge.

Foreign students came to Uppsala to study with Svedberg, and an intense period of research began. Proteins were prepared, purified and studied in the ultracentrifuge. By the end of the 1920´s some fifteen different proteins had been studied by Svedberg and his collaborators. Quite unexpectedly it was found that most of the proteins were monodisperse, that is, they had constant and well-defined molecular weights.

According to Svedberg´s own statement, the greatest sensation occurred when the blue haemocyanin from the blood of the vineyard snail, Helix pomatia, was centrifuged. According to the copper content of the blood, the protein should have a minimum molecular weight of 15,000 to 17,000. This would mean that there should be only a very small difference between the protein concentration at the top and the bottom of the cell at sedimentation equilibrium. After centrifugation for a short time, it was found that the haemocyanin sedimented with a razor-sharp boundary down through the cell. A preliminary estimate showed that the molecular weight had to be in the millions and that all the molecules had to be of the same size. It was the first time that such giant molecules had been observed. Later on Svedberg could show that this haemocyanin had a constant molecular weight in buffers within a broad pH-region.

Gradually Svedberg came to the conclusion that certain rules existed for the molecular weights of the proteins. In a "letter to the editor" published in Nature, June 8, 1929, he put forward the so-called multiple hypothesis. He wrote: "Our work has been rewarded by the discovery of a most unexpected and striking relationship between the mass of the molecules of the same protein at different acidities, as well as of a relation concerning the size and shape of the protein molecules".

"It has been found that all stable native proteins so far studied can, with regard to molecular mass, be divided into two large groups: the haemocyanins with molecular weights of the order of millions and all other proteins with molecular weights from about 35,000 to about 210,000".

In order to explain the regularities, Svedberg continues: "When looking for an explanation of these unexpected regularities, it would be well to bear in mind the fact already brought out by many biochemical experiences, namely that Nature in the production of organic substance within the living cell seems to work only along a very limited number of mainlines. The great variety appears in the specialization of details. Thus, it would seem that the numerous proteins are built up according to some general plan which secures for them only a very limited number of different molecular masses and sizes when present in aqueous solutions. By varying the constituents of the different proteins (different percentages of different amino-acids, etc.) the chemical and electrochemical properties may be varied sufficiently to enable the cells to make use of them for their different purposes".

These results and lines of thought were so new and sensational that several chemists doubted the reliability of the centrifugal method, especially after Svedberg´s introduction of his hypothesis of the multiple system for the molecular weights of the proteins.

Svedberg´s view of the proteins before the exciting results with the haemoglobin is very clear from a letter to Professor Edwin J. Cohn at Harvard University on May 30, 1926:

"My dear Dr. Cohn,

The determination of the molecular weight claims most of our interest — as you may imagine. A young American, Mr. J.B. Nichols, who was working with me during my stay at Wisconsin University, has been working in my lab during the last year on the mol. weight of egg albumin. As you perhaps remember, the value of the centrifugal method is not only that it gives a more reliable means of determining the mol. weight than does the osmotic method, but it further gives a means of ascertaining whether a certain protein solution is built up of molecules of equal weight or not. I had always been of the opinion that the protein solutions were rather ill defined systems containing particles of all kinds of sizes - the so-called mol. weight only being a mean value - and I was therefore very much astonished when the centrifugal method showed that the haemoglobin solutions were built up of special-sized well defined molecules. Now Mr. Nichols in his study of the egg albumin (we use the ultraviolet range 280-330 nm) at first found a drift of the mol. weight from about 33,000 to 53,000 indicating that the egg albumin was not a simple substance. Repeated purifying by means of crystallization and electrodialysis, however, gave the result that with increasing purity the drift in the values became less and less. Finally we reached a constant value very close to 34,000! We have also taken up the study of ovonucoid, but have not finished that protein yet. The proteins which we are preparing for work next term are edestin, phycocyan, phycoerythrin, haemocyanin, serumalbumin and globulin. It is our intention to take up in turn the study of the mol. weight of most of the well defined and important proteins. There is a second possible method of determining the mol. weight by centrifuging as I have already pointed out in the Zsigmondy-Festschrift, namely by measuring the velocity of sedimentation. This method, however, requires a very powerful ultracentrifuge. We have constructed and finished the installation of such machinery which can give us up to 90,000 times gravity (the ultracentrifuge we used before can only give 7,000 times gravity). It would give me great pleasure to show you this apparatus".

After Svedberg had got the 1926 Nobel Prize, it was not difficult to get the Parliament to grant funds for a new institute of physical chemistry at Uppsala University. At the beginning of 1931, Svedberg could move over to his new institute of physical chemistry. A new and reconstructed ultracentrifuge, giving a centrifugal field twice as intense as the first oil turbine ultracentrifuge, had also been installed there.

In order to increase the efficiency of the new ultracentrifuge, a number of different rotors were constructed and tested. The size and shapes of these rotors were designed and calculated in collaboration with Gustaf Boestad, later professor at Royal Institute of Technology (KTH) in Stockholm. The different rotors were tested with varying success before the prototype of the final type of rotor was tested in January 1939. Since 1931 twenty different rotors had been tested. Nine of these rotors had burst or exploded during the test runs.

The testing of the new rotors was very stressful for Svedberg, and he used to say that each test run would shorten his life by one year! However, he actually lived to pass his 86th birthday.

There is no doubt that unsuccessful test runs and other difficulties, such as vibrations in the centrifuges, were very distressing for him, and he was sometimes not far from giving up further work on the development of the centrifuge and concentrating on other problems. His deep interest in proteins and his eagerness to prove or disprove his hypothesis of the multiple system for the molecular weights of proteins made him continue to develop the ultracentrifuge. After having left Uppsala at the end of his sixties, he said that the 1930s were probably the happiest years of his scientific life.

On November 17, 1938, the Royal Society of London arranged a discussion on "The Protein Molecule" (ref.3) with Svedberg as the main speaker. He began his speech by saying: ´The proposal of the subject for this discussion is in itself a remarkable thing and a symbol of the spirit of this meeting. A few years ago, the proposal would have looked preposterous. Proteins were known as a mysterious sort of colloid, the molecules of which eluded our search. What is it then that has happened in these years? Why is the most distinguished scientific society of this country inviting a discussion on the protein molecule?´ Somewhat later in his lecture, Svedberg said: ´Investigations along different lines have given the result that the proteins are built up of particles possessing the hall-mark of individuality and therefore are in reality giant molecules. We have reason to believe that the particles in protein solutions and protein crystals are built up according to a plan which makes every atom indispensable for the completion of that structure´.

This meeting at the Royal Society in London was in a way the climax of the protein epoch in Svedberg´s life. A few months later the final rotor type for the oil turbine ultracentrifuge was successfully tested, and at the same time the manuscripts for the monograph "The Ultracentrifuge" (ref.4) and "Die Ultrazentrifuge" were sent to the British and German publishers, respectively.

The result of the protein epoch may be summarized as follows: the introduction of the ultracentrifuge method led to the discovery that proteins were well defined chemical substances with a fixed molecular weight. The sedimentation velocity method and later Tiselius´ electrophoretical method made it possible in a much more direct way than earlier to follow the isolation and purification of the individual proteins and to show when it was satisfactory. It let Svedberg put forward his multiple hypothesis for the molecular weights of the proteins, and even if this hypothesis was by no means of a general nature, it meant very much to the development of protein chemistry, especially during the 1930s and the beginning of the 1940s. It started a new interest in the proteins. Chemists, biologists, and physicists discovered that the proteins no longer could be regarded as ill-defined lyophilic colloids. On the contrary, they were well-defined, interesting and very important substances, well worth studying. The door had just been opened a little to the new science of molecular biology (ref.5).

Shortly before the outbreak of World War II, Svedberg and his co-workers had started to study the effect of different types of radiation, such as ultraviolet light, α-particles, γ-rays, neutrons, and ultrasonics on proteins. For the production of the neutrons a small neutron generator was built. It was even used for a short time for the production of some radioactive isotopes for medical studies, but was inadequate for any major production of radioactive isotopes.

Svedberg's exceptional working capacity and passion for research were very infectious. His example and his interest in his co-workers and students made the working conditions at his institute very pleasant. We who had the good fortune of working together with him look back to this period with much gratitude.

REFERENCES

1 T. Svedberg and R. Fåhraeus, A New Method for Determination of the Molecular Weight of the Protein, J. Am. Chem. Soc., **48**, 430–438 (1926).
2 A. Tiselius and S. Claesson, The Svedberg and Fifty Years of Physical Chemistry in Sweden. Ann. Rev. Phys. Chem., **18**, 1–8 (1967).
3 T. Svedberg, A Discussion on the Protein Molecule. Proc. Roy. Soc. (London) A, **170**, 40–56 (1939).
4 T. Svedberg and K.O. Pedersen, The Ultracentrifuge. Clarendon Press, Oxford (1940). Reprinted by Johnson Reprint Corp., New York (1959).
5 K.O. Pedersen, The Svedberg and Arne Tiselius. The Early Development of Modern Protein Chemistry at Uppsala. In G. Semenza (Ed.), Selected Topics in the History of Biochemistry. Personal Recollections. Comprehensive Biochemistry, Vol. **35**, Chapt. 8, p. 233–281. Elsevier Science Pub. (1983).

6. Svedberg and the Polysaccharides

Ingvar Jullander

Swedish Pulp and Paper Association, Villag. 1, 114 32 Stockholm, Sweden

Abstract — The main part of Svedberg´s work on carbohydrates was
published in the years 1938 - 1952. Sedimentation velocity experiments
seldom revealed the existence of molecules with distinctly different
molecular mass, i.e. separate peaks in the sedimentation diagrams.
Determination, directly or indirectly, of the frequency function of the
molecular mass was a main problem. It was approached from different
angles as indicated in the review.

INTRODUCTION

The carbohydrate based research on properties of macromolecules started in earnest towards
the end of the 1930s, i.e. later than the protein research dealt with in the previous paper
by Kai Pedersen.

The research summarized below was published mainly between 1938 and 1952, much of it is
included in doctoral theses. The main coworkers of Svedberg working on carbohydrates during
this period were:

> Nils Gralén
> Hans Mosimann
> Bengt Rånby
> Sigurd Säverborn
> and myself.

Unless specifically mentioned, research results to be mentioned were obtained in cooperation
with these five scientists. It is only fair to add, though, that the first ultracentrifugal
experiments with cellulose had already been published in 1930 by Alfred Stamm, USA (ref.1)
who spent some time at the Institute. The name of E.O. Kraemer, also from the USA should be
added.

Had not the intuition of Svedberg led him to start his macromolecular investigations
with reasonably simple protein systems, mono- or pauci-disperse, the ultracentrifuge may
never have been developed into a precision instrument. To get reliable results from
polymolecular systems like cellulose turned out to require a very advanced experimental
technique.

When the author started working for The Svedberg in the autumn of 1939 the oil turbine
ultracentrifuge was highly developed, running smoothly and giving a first class technical
service. To put it briefly and using an expression from the process industry of today: the
availability was very high indeed!

SPECIAL INVESTIGATIONS

Providing a contrast from proteins, work on sedimentation pictures of juices from bulbs of
different species of lilies are presented in Fig. 1. The juices contain polysaccharides
held in storage by the plants. The sedimentation was observed by the scale line displacement method developed by Ole Lamm (ref.2). The position of a peak along the x-axis gives
the sedimentation constant characterizing a macromolecule. The areas under a peak is
proportional to the concentration of that macromolecule. There are two or in one case three
types of macromolecule present. In some cases it could be ascertained that the faster
sedimenting peak contained a protein, the slower a carbohydrate.
Altogether no less than about 75 different species were analysed, their sedimentation
diagrams differed widely. In two cases the average molecular masses of the polysaccharides
were determined, they were found to be 20,000 and 50,000 respectively.

Only seldom did sedimentation experiments on polysaccharide material give separate
peaks like those just shown. However, water extractable material from the wood of the larch
tree (Larix) turned out - quite unexpectedly - to consist of two polysaccharides of widely
different molecular mass, Fig. 2. This is illustrated by sedimentation diagrams from 10 to
110 minutes after reaching full centrifugation speed. The α-component, probably an araban,
was monodisperse with a molecular mass 16,000; the β-component, probably a galactan, had a
molecular mass 100,000 and is considered also to be more or less monodisperse.

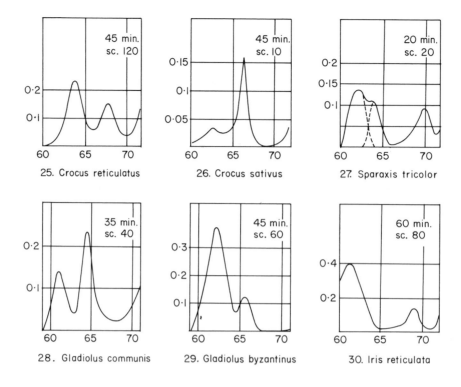

Fig. 1. Sedimentation diagrams of juices from bulbs of lilies.
Abscissa: distance (mm) from centre of rotation, ordinate: scale line
displacement (mm). Centrifugal force 300,000 - 350,000 $\underline{g}$. From ref.3.

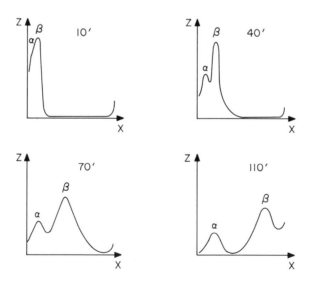

Fig. 2. Sedimentation diagrams of water soluble polysaccharides from
the wood of the larch. Centrifugal force around 250,000 $\underline{g}$. $\underline{X}$ =
distance from centre of rotation, $\underline{Z}$ = scale line displacement. From
ref.4.

Sedimentation experiments on solutions obtained through extraction of wood or dissolution of
holocellulose in the classical cellulose solvent, cuprammonium, also gave two peaks,
presumably due to hemicellulose and cellulose.
    Crosslinking of nitrocellulose in solution with titanium tetrachloride gave a
sedimentation diagram with three peaks. The rate of sedimentation of the fastest peak - the

γ-peak - slows down to almost a tenth of its speed at the start when approaching the bottom of the centrifuge cell. This is a characteristic behaviour of gels, the loose net structure being compressed like an elastic spring. The α-peak contains single molecules not cross-linked and the β-peak is assumed to consist of double, and possibly triple or more, molecules.

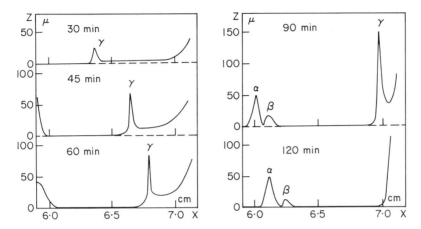

Fig. 3. Sedimentation diagram of nitrocellulose, 0.4 %, dissolved in butyl acetate to which 0.5 equivalents of titanium tetrachloride has been added. Centrifugal force 170,000 $\underline{g}$, $\underline{X}$ = distance from centre of rotation, $\underline{Z}$ = scale line displacement. From ref.5.

A special study was devoted to polyuronides, i.e. carbohydrates with carboxyl groups. Many types exist in nature. This study concentrated on pectic substances, alginic acid and gum arabic.

The molecular masses varied from 50,000 to 500,000 as measured by sedimentation and diffusion. No relation was found between the jellying power of pectins and the molecular mass. The necessity of making measurements of the molecular properties of these poly-electrolytes in salt solution to avoid electric charge effects was clearly demonstrated. Sedimentation, diffusion and viscosity determinations gave peculiar results in pure water.

CELLULOSE AND CELLULOSE DERIVATIVES

As already mentioned most of the work was done on cellulose and cellulose derivatives, especially nitrocellulose[*]. The polydispersity or polymolecularity was a mutual problem in all these investigations. It was at the same time a challenge and a headache to find the frequency function of the molecular mass.

This review concentrates on different aspects of this problem instead of giving detailed results from varying types of cellulose and its derivatives.

Polymolecularity in principle

Knowledge of a well defined average molecular mass plus knowledge of the shape of the frequency distribution around that average gives complete information on the poly-molecularity. This was the desirable but seldom obtainable, ultimate aim.

There are different types of averages:

Number average $\qquad \overline{\underline{M}}_{\underline{n}} = \dfrac{\int d\underline{c}}{\int d\underline{c}/\underline{M}}$

Weight average $\qquad \overline{\underline{M}}_{\underline{w}} = \dfrac{\int \underline{M}\, d\underline{c}}{\int d\underline{c}}$

Z-average $\qquad \overline{\underline{M}}_{\underline{z}} = \dfrac{\int \underline{M}^2 d\underline{c}}{\int \underline{M}\, d\underline{c}}$

where $\underline{M}$ = molecular mass, $\underline{c}$ = concentration and $\overline{\underline{M}}_{\underline{n}} < \overline{\underline{M}}_{\underline{w}} < \overline{\underline{M}}_{\underline{z}}$.

---

[*] "Nitrocellulose" is a technical and commercial name for cellulose nitrate.

From osmotic pressure measurements, number average molecular masses are obtained, through sedimentation equilibrium experiments, weight and Z-averages are found. In the latter case solutions of macromolecules are centrifuged at lower rotational speeds compared with sedimentation velocity experiments, but longer times until equilibrium between sedimentation and diffusion has been reached. Viscosity measurements give weight averages provided Staudinger's rule is applicable and the calibration correctly made.

Molecular mass is also determined from sedimentation velocity plus diffusion using Svedberg's formula:

$$\underline{M} = \frac{\underline{s}}{\underline{D}} \quad \frac{\underline{RT}}{(1 - \underline{V}\rho)}$$

where $\underline{s}$ and $\underline{D}$ are sedimentation and diffusion constants
$\underline{R}$ is the gas constant
$\underline{T}$ is temperature
$\underline{V}$ is the partial specific volume of the macromolecule and
ρ is the density of the solution.

A polymolecular substance has a frequency distribution both in $\underline{s}$ and $\underline{D}$. It is possible to measure weight averages for both of them but substitution of these averages in Svedberg's formula unfortunately does not give a weight average molecular mass. Calculations showed that the average obtained usually lies between the number and weight averages and varies with the shape of the distribution curve.

In principle measurements of molecular properties on linear macromolecules should be made at infinite dilution. In practice this means experiments at varying, low concentrations and extrapolation to concentration zero. This made it necessary to refine the experimental techniques as much as possible.

Osmotic pressure

For measurements of number averages a new type of osmometer, an osmotic balance, was constructed, later considerably improved by Bertil Enoksson, Fig. 4. The height difference at equilibrium between solvent and solution, corresponding to the osmotic pressure was weighed instead of directly measured as a length. In nitrocellulose solutions it was in this way possible to reach dilutions corresponding to about one tenth of previously published osmotic pressures.

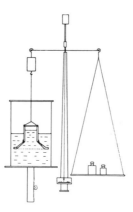

Fig. 4. Principle of the osmotic balance. From ref.6.

Sedimentation velocity

The oil turbine ultracentrifuge was, as already mentioned, working admirably. When investigating nitrocellulose dissolved in acetone it was possible to make meaningful measurements down to 0.01-0.02 % concentration. In this range a temperature difference of $0.05^{o}C$ in the centrifuge cell would have caused a complete heat convection.

Diffusion

From Svedberg's formula it is clear that measuring the diffusion is just as important as measuring the sedimentation velocity in providing the basis for molecular mass determinations. The diffusion constant was determined by carefully stratifying solvent above

solution and then for a few days – sometimes a week or slightly more – studying the blurring
of the boundary by the Lamm scale method.

Fig. 5 illustrates the diffusion of flax fibre cellulose dissolved in cuprammonium.
Different exposures have been recalculated to the same time, the result indicates high
experimental precision. The diffusion constant increases considerably with concentration,
causing the skewness of the curve – solution to the right, solvent to the left. A weight
average diffusion constant can be calculated, but the subsequent extrapolation to zero
concentration is often large and introduces a considerable uncertainty. In spite of much
work both theoretically and experimentally on the diffusion technique it seems clear that
the diffusion experiments were the weakest link of the chain.

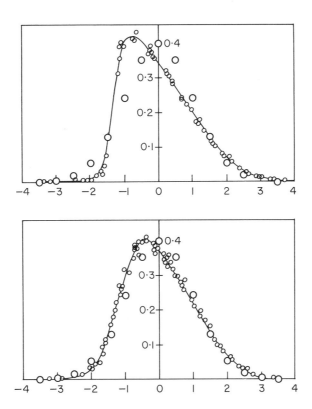

Fig. 5. Flax fibre cellulose dissolved in cuprammonium, concentration
0.20 % (upper curve), 0.10 % (lower curve). Diffusion curves in normal
coordinates recalculated from exposures at different times. The large
circles indicate the position of a corresponding normal curve. From
ref.7.

Polymolecularity in practice

Degrees of polymerization for cellulose from sedimentation velocity and diffusion around and
above 10,000 were reported in some instances. These values were later considered unlikely
to be so high because of too low values for the extrapolated diffusion constants. A number
of different measures for polymolecularity were proposed and used.

- From sedimentation velocity alone the broadening of the sedimentation peak with time was
  determined at different concentrations and then extrapolated to infinite dilution.
- From diffusion experiments alone diffusion constants were determined by the so-called
  area method and the moment method. The last gives a weight average, the former a lower
  average. The quotient between them can be used as a measure of polymolecularity.
- The quotient between any two of the differently defined measures of average molecular
  mass is a relative measure of polymolecularity.
- If a logarithmic distribution function is assumed – this is in itself not unreasonable –
  knowledge of the quotient $\underline{M}_w/\underline{M}_n$ is sufficient to define the form of the distribution
  curve.
- It was also possible under certain assumptions to transform sedimentation pictures of
  nitrocelluloses to corresponding frequency distributions of the molecular mass.
  Diffusion is negligible during the short time of a sedimentation run.

- The universally, and at that time most often used method in the study of polymolecularity was fractionation, which is tedious and time consuming. It also has a weakness in that the polymolecularity of the single fractions is unknown.
- Fractionation of nitrocellulose combined with sedimentation velocity experiments and viscosity determinations on the fractions was found to give detailed information on the polymolecularity. The broadening of the sedimentation pictures of the separate fractions was used as a relative measure of their polymolecularity. The frequency distributions of single fractions were taken as triangles, the width of which were proportional to the broadening. A summation gave detailed diagrams, one of which is shown in Fig. 6.

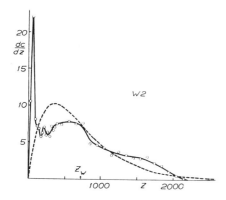

Fig. 6. Frequency curve for the degree of polymerization ($\underline{Z}$) of nitrated bleached sulphate pulp (rayon grade). The dotted line indicates the theoretical frequency curve resulting from a homogeneous depolymerization of a very long polymer chain ($\underline{Z}$ = 100,000). From ref.8.

Supermolecular structure

So far the review has dealt with studies of cellulose fibres on the molecular scale. At the other end of the scale fibres have always been carefully investigated with the microscope. To bridge the gap in between, the transmission electron microscope appeared on the scene towards the end of the 1930s. It has later been to a large extent replaced by the scanning electron microscope.

Svedberg was able to acquire, shortly before the war, an electron microscope made by Siemens in Germany. This was used for an extensive study of the supermolecular (submicroscopic) structure of cellulose fibres of very different origins. The investigations were supplemented by X-ray studies of the crystallinity of cellulose. Not only wood and cotton but also cellulose from algae, bacteria and animals were studied. In fact mantle animals, tunicates, produce cellulose called tunicin. The great similarity between this tunicin and cellulose had already been pointed out in 1845.

Disintegration of cellulose material in water suspension with ultrasound revealed the existence of microfibrils also called micelle strings, Fig. 7. Independent of the origin, the diameter of these strings was about the same, 8 - 11 nm.

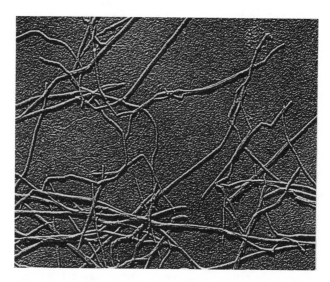

Fig. 7. Micelle strings from tunicin (animal cellulose). Magnification 51,000. From ref.9.

They differed, however, in their behaviour in mercerization. The soda lye concentration necessary to obtain a transformation in the crystal structure from cellulose I to cellulose II increases in the order: wood, cotton, bacterial cellulose, animal cellulose, Fig. 8. The water sorption from moist air followed the same general trend.

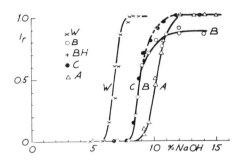

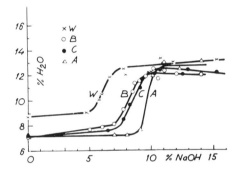

Fig. 8. Phase transition (upper plot) from cellulose I to cellulose II by treatment with alkali. Wood (W), cotton (C), bacterial (B, BH) and animal cellulose (A). Lower plot: water sorption at 65 % rel. humidity. From ref.8, p.8.

When treated with boiling sulphuric acid the micelle strings are broken down to micelles (Fig. 9) which unexpectedly show hydrophobic properties. An approximate calculation shows that one micelle string will contain about 240 cellulose chains. A cross section of a softwood tracheid will in turn contain about 5 million such micelle strings.

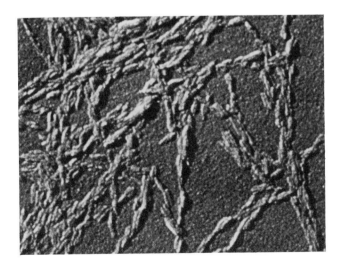

Fig. 9. Micelles from mercerized wood cellulose. Magnification 85,000. From ref.10.

CONTINUED WORK

It should be added that investigations on the molecular properties of cellulose and
cellulose derivatives were continued under the leadership of Stig Claesson, the successor of
The Svedberg. New theories, new and improved instrumentation were developed and several
cellulose derivatives were thoroughly studied such as cellulose xanthate, hydroxyethyl
cellulose, ethyl-hydroxyethyl cellulose, carboxymethyl cellulose and, as examples of
cellulose derivatives with large substituents, cellulose tricarbanilate and cellulose
tributyrate. This is all outside the scope of this presentation.

REFERENCES

1.   A.J. Stamm, J. Am. Chem. Soc., 52, 3047, 3062 (1930).
2.   O. Lamm, in The Ultracentrifuge (eds. T. Svedberg and K.O. Pedersen), p. 289, Clarendon
      Press, Oxford (1940).
3.   N. Gralén and T. Svedberg, Biochem. J., 34, 234 (1940).
4.   H. Mosimann and T. Svedberg, Kolloid-Z., 100, 100 (1942).
5.   I. Jullander, Studies on Nitrocellulose, Arkiv Kemi Min. Geol, 21A, No. 8, p. 132
      (1945).
6.   B. Enoksson, J. polym. Sci., 3, 314 (1948).
7.   N. Gralén, Sedimentation and Diffusion Measurements on Cellulose and Cellulose
      Derivatives, Dissertation, p. 68. Uppsala (1944).
8.   B.G. Rånby, Fine Structure and Reactions of Native Cellulose, Dissertation, p. 13,
      Stockholm (1952).
9.   B.G. Rånby, Arkiv Kemi, 4 (no. 13), 249 (1952).
10.   B.G. Rånby, Acta Chem. Scand., 6, 130 (1952).

7. Svedberg and the Synthetic Polymers

Per-Olof Kinell

Department of Physical Chemistry, Umeå University, S-901 87 Umeå, Sweden

Abstract - A survey is made of the work done in The Svedberg´s laboratory on polystyrene, polychloroprene and poly(methyl methacrylate). Results on molecular properties and molar mass distributions are exemplified. Synthesis of polychloroprene was part of a programme for a domestic production of rubber during World War II. Some personal remembrances of The Svedberg the scientist and the man are given.

INTRODUCTION

On April 24th 1942 The Svedberg gave a short communication for the Swedish Broadcasting Corporation from which I quote the following lines (ref.1):
"Thanks to the enthusiasm and energy with which my collaborators have set to work we have during the last few days been able to produce raw rubber in Uppsala and send samples to a factory for technical tests. At a meeting the other day with the Sveriges Kemiska Industrikontor the first articles made from Swedish synthetic rubber by the laboratory of Helsingborgs Gummifabrik were demonstrated. It is to be hoped that only a short time will elapse before our chemical industry will be able to exchange the miniature boot, which was exhibited on that occasion, for a similar boot but of normal size."
   The background to Svedberg´s communication was the serious problem that we had in Sweden during the first years of World War II in meeting the urgent demand for rubber for our military units. The possibilities for importing rubber were severely restricted. In order to handle the difficulties the Governmental Commission on Industry requested in November 1941 The Svedberg to use his laboratory to find a process for production of the synthetic rubber Neoprene based on Swedish raw materials. This task had to be performed in the shortest possible time.
   This commission given to Svedberg involved a widening of the research programme at the Institute of Physical Chemistry. So far no work had been done on the formation of macromolecules. All investigations were focussed on the properties of naturally occurring large molecules, viz. proteins, cellulose and water-soluble polysaccharides. Only some polymer analogous transformations had been performed, e.g., to obtain cellulose derivatives. From now on the complete synthesis of macromolecules from their building stones became an integrated part of the research activities. This also included the formation of the building stones themselves. Furthermore the ultracentrifuge became a tool not only to investigate molecular size and shape per se but also to obtain control so that molecules of a suitable size would be formed in the polymerization process. In addition several new techniques had to be developed, e.g., to study latexes, elastic behaviour and swelling in organic solvents.
   The achievements made in the field of synthetic polymers under the guidance of Svedberg were not only concerned with elastomeric materials but also included important work on polystyrene and poly(methyl methacrylate). In the following I will give a chronological review and present some of the most important results obtained.

POLYSTYRENE

In the beginning of the 1930s there was a lively discussion about the concept of macromolecules mainly based on the work and propositions made by Staudinger. The debate also included the applicability of various methods for determination of molecular size because a knowledge of this parameter would be a conclusive proof of the existence of threadlike macromolecules in which the repeating units were linked together by covalent bonds. Staudinger (ref.2) expressed some doubt about the use of the osmotic method because of the low osmotic pressures: a 0.023 mM solution of polystyrene of molar mass 100,000 gives a pressure of 53 Pa. He was also critical of diffusion measurements. About the use of sedimentation in centrifugal fields he stated (ref.2): "die Methode von Svedberg, die Teilchengrösse von Kolloiden mittels der Ultrazentrifuge zu bestimmen, dürfte zur Bestimmung des Molekulargewichts hochmolekularer Stoffe, deren Teilchen Fadenmoleküle sind, ebenfalls nur mit Vorsicht angewandt werden." The reason for this view was mainly based on the opinion that threadlike molecules due to their large surface area would exhibit such a high frictional resistance that the movement of these molecules would be extremely slow even in very strong centrifugal fields. Furthermore the velocity would depend upon the orientation of the molecules with respect to the field.

   Hoping to clarify the situation, Signer and Gross in October 1932 began at Svedberg's laboratory an investigation into the sedimentation behaviour of various polystyrene samples prepared and viscosimetrically characterized by the Staudinger school. The results of this work were collected in three papers; the first one treating the sedimentation velocity properties of polystyrene (ref.3), the second reporting molecular mass determinations from equilibrium sedimentation (ref.4) and the third exploring the ultracentrifugal separability of polystyrene fractions (ref.5). It has to be mentioned that the experimental work necessitated the mastering of some technical problems in making ultracentrifugal cells tight for organic solvents. Furthermore precautions had to be taken to prevent evaporation of the liquid in order to avoid convection currents.

   Signer and Gross found that in one and the same solvent the polystyrene molecules could easily be centrifuged, the same values of the sedimentation constant were obtained over a range of sedimentation distances, the sedimentation constant did not depend on the centrifugal force and exhibited distinctly different values for the various samples. The concentration dependence became more pronounced the higher the molecular mass. From a series of measurements in good solvents they concluded that the polystyrene molecules were present as free, highly solvated entities. Results from measurements in bad solvents indicated a more compact and less solvated molecular state. Some data compiled from ref.3 are collected in Table 1. The concentration dependence of the sedimentation constant, S, has been characterized by the parameter k in the expression $S = S_o/(1 + kc)$.

TABLE 1. Sedimentation data of polystyrene samples

M_V	S_o/s	k/M^{-1}
15 000	3.8	4.8
31 000	6.7	12
135 000	16.0	52
260 000	25.4	123
600 000	38.1	220

In equilibrium sedimentation (ref.4) large deviations from ideal behaviour showed up in very dilute solutions. Establishment of the equilibrium was very slow (four weeks for a sample with molar mass 270,000) due to the low diffusion constant of the threadlike molecules.

   At sufficiently low concentrations polystyrene samples having different molar masses sedimented independently of each other, e.g. with M = 1,100,000 and 270,000 at c = 0.002 M and M = 270,000 and 80,000 at c = 0.012 M (ref.5). Furthermore a solvent with a low density compared to that of the polymer and a suitable rotor speed gave the high sedimentation velocity which was necessary to suppress the influence of diffusion on the concentration distribution of the polymer masses in the cell. Then the concentration gradient curve gave a true picture of the polymolecularity of the samples.

   On this occasion I would like to give my compliments to Professor Signer for the performance of these measurements. They have become of utmost importance as a foundation for all later ultracentrifugal studies of synthetic polymers.

POLYCHLOROPRENE

Polymerization

The decision taken by the Governmental Commission on Industry to start domestic production of synthetic rubber of the Neoprene type was based mainly on the fact that the situation with respect to the raw materials was favourable. We already had a factory for producing calcium carbide, also hydrogen and chlorine were available. The alternative discussed was the Buna rubber developed in Germany. Ethyl alcohol, which could be converted into butadiene, was easily available from the cellulose industry. However, this possibility was rejected, one of the reasons being our delicate political relations with Germany. The year was 1941.

   The work in Uppsala started under the leadership of Professor Sven Brohult immediately after the commission had been given. The first task became to go over relevant publications and patents. Often this caused serious problems because due to the war we were almost completely cut off from foreign institutions and libraries. Most useful as a source of information was the work done at the E.I. duPont Company and some Russian work on Sovprene. The main activities undertaken included synthesis of chloroprene, polymerization in bulk and in emulsion, study of the polychloroprene molecules in solution (viscosity, sedimentation, diffusion, fractionation, adsorption analysis) and exploration of bulk polychloroprene (elasticity, crystallization, swelling). A report on the work performed up to 1946 is given by Svedberg and Kinell (ref.6). Some of the results are also presented in ref.7-10.

The main steps in the formation of polychloroprene are the following:

1. Formation of acetylene

$$CaC_2 + H_2O \rightarrow CH\equiv CH + CaO$$

2. Conversion of acetylene into monovinylacetylene

$$2CH\equiv CH \rightarrow CH_2=CH-C\equiv CH$$

3. Formation of chloroprene

$$CH_2=CH-C\equiv CH + HCl \rightarrow CH_2=C.Cl-CH=CH_2$$

4. Polymerization of chloroprene

$$CH_2=C.Cl-CH=CH_2 \rightarrow -(-CH_2-C.Cl=CH-CH_2-)\overline{_n} \quad .$$

I will not dwell very much upon the work done on the synthesis of monovinylacetylene and chloroprene. Various catalyst recipes based on copper(I) chloride were tried in order to get high yields and to reduce the formation of by-products, e.g., divinylacetylene. Exploratory experiments were made on a laboratory scale. When useful processes had been developed a pilot plant was erected in one of the corridors of the laboratory. This was ready to be used in the autumn of 1942.

In parallel with the production of chloroprene, attempts were made to polymerize it. The first experiments in December 1941 gave only 1-2 g h^{-1} but in February 1942 the production rate could be increased to 60 g h^{-1}. It was then successively increased as larger quantities of chloroprene became available. The goal for the pilot plant scale was set at 10-20 g h^{-1}.

The first technique used was polymerization in emulsion with ammonium oleate as an emulsifying agent. The combination tetralin peroxide/p-nitroaniline as a catalyst and phenyl-α-naphthylamine as a stabilizer was used. The first successful result was achieved in January 1943 with the polymerizate Ep 39, which was obtained in a quantity of 165 g corresponding to a yield of 66 per cent. Its mechanical properties were compared with those of Neoprene E at the rubber factory in Helsingborg with the results given in Table 2.

TABLE 2. Mechanical properties of polychloroprenes

Sample	Strength at rupture (kg cm^{-2})	Elongation at rupture (per cent)	Hardness ($^{\circ}$Shore)
Ep 39	198	670	64
Neoprene E	154	680	65

Some of the difficulties encountered were control of temperature, influence of copper impurities, balancing of α-/μ-polymer ratio and preventing formation of ω-polymer. Most of the work was done at the specially arranged Swedish Rubber Research Laboratory on the outskirts of Uppsala, where polymerizations on a pilot level could be made. An urgent need for larger amounts of rubber to be supplied to the Swedish rubber factories called for the planning of a technical scale rubber production. This resulted at midsummer in 1944 in the start of a plant at Ljungaverk in northern Sweden. On a later occasion we had the pleasure of demonstrating this plant to Dr Burton Nichols as a representative of the E.I. duPont Co.

Some experiments were also performed with photo and bulk polymerization. The results were not too promising. Instead the so called belt method was developed. Chloroprene was polymerized into a syrup containing 5-10 % polymer. This syrup was continuously spread out in a layer on a thin conveyer belt made of stainless steel. The belt passed through a steam chamber where excess chloroprene evaporated. Finally the polymer layer was removed from the belt. The method was successfully tried in full·scale operation. It gave polymerizates of a more uniform quality compared to those from emulsion polymerization.

A number of copolymers between chloroprene and various acrylic compounds were produced using the emulsion technique. This supplied a demand for a rubber quality more resistant to oil and petrol than the pure polychloroprenes.

Quite comprehensive studies were performed on unpolymerized and polymerized latexes, e.g., particle size distribution, stability and electrophoresis (ref.6). Also the size and shape of potassium myristate and laurate micelles were investigated in media of various ionic strengths (ref.9). The heat of polymerization of chloroprene was measured in an isothermal calorimeter at 334.5 K. The values obtained increased slightly with the amount of initiator (benzoyl peroxide). Extrapolation to zero concentration gave 67.8 kJ per monomer unit (ref.10).

Properties of polychloroprene molecules

The original object of the studies on polychloroprene molecules in solution was to characterize the polymerizates obtained under different conditions both on a laboratory and an enlarged scale. Also the ageing of the materials was of importance to study. In all some 250 rubber samples had been studied up to 1946. One of the aims was to compare the polychloroprenes obtained in emulsion, A-polymers, with those from bulk polymerization, B-polymers. Of the B-polymers 80 per cent of the samples were completely soluble in chloroform compared to only 30 per cent of the A-polymers. This indicates a higher degree of crosslinking in the latter case. The viscosity number of the soluble samples varied from 50 to 400 cm^3g^{-1} for both types of polymer. The majority of the soluble fractions of the partially soluble samples had viscosity numbers less than 150 cm^3g^{-1} (ref.6).

The sedimentation diagrams of the polychloroprenes in chloroform solution always showed one single peak which, however, became very low and wide in dilute solutions indicating a large range of molecular sizes in every sample (refs.6, 7). The sedimentation constants covered the interval 2-6 s. Most of the B-polymers had a value of about 3.5 s. The concentration dependence of the sedimentation of the A-polymers was greater than that of the B-polymers. Also diffusion constants were determined and when combined with the sedimentation constants the molar masses were obtained. For both types of polymer these ranged from 50,000 to about 1,000,000. A Neoprene sample gave 3.3 s and the molar mass 204,000. The conclusions drawn from these measurements were that the A-polymers had a higher polymolecularity than the B-polymers and that the B-polymers consisted of threadlike macromolecules which were more freely drained than those of the A-polymers.

From the molar masses and the viscosity values it was not possible to determine any K- and a-values in the Mark Houwink equation. The reason was judged to be the great polymolecularity of the samples. However, fractionated samples of Neoprene E the molar masses of which were determined from equilibrium sedimentation in amylacetate solutions gave the values $1.0x10^{-2}$ cm^3g^{-1} and 0.80, respectively (ref.6). The same polymer in toluene gives $5.0x10^{-2}$ cm^3g^{-1} and 0.62 (ref.11).

Frontal analysis on active carbon of some polychloroprene samples gave diagrams with at least two steps, indicating a separation of the material in fractions. Examination of the first two fractions in the ultracentrifuge resulted in slightly higher S-values for the fractions compared to the original sample, e.g., 5.9, 5.8 and 5.3 s (ref.6).

Elasticity

Unvulcanized polychloroprene samples were studied with respect to relaxation, after rapid stretching, and to elastic and thermal recovery (ref.6). Relaxation to a constant stress value was completed within less than 60 min. for all samples. The relative stress reduction became lower the higher the original elongation, a behaviour which was assumed to be connected with alignment of the molecular chains and crystallization. The elastic recovery was almost complete after thermal treatment in steps from 18 up to 50°C. The recovery was found to be most efficient at 35°C, i.e., at the melting point of the crystallites formed during stretching (cf. ref.12).

Stress strain curves for unvulcanized samples exhibited a high resistance towards deformation at low strains (ref.8). Except for this effect the curves behaved in a similar fashion as for vulcanized rubber, i.e., a rather slow increase in stress at middle strains followed by a more rapid rise at higher strains. The hysteresis losses were found to be about 30 per cent. A decrease with temperature was observed.

Some studies were also made to apply the molecular theory of rubber elasticity to the stress strain behaviour of unvulcanized polychloroprene. The results were, however, not too successful due to a large change in internal energy with elongation in addition to the change in entropy (ref.8).

Finally experiments on dynamic deformation of rubber were initiated. However, before 1946 the set up had only been tested on vulcanized Neoprenes. These exhibited, at a frequency of 1100 r.p.m., two critical temperatures for the change in deformation, at -35 and +30°C, respectively (ref.6).

X-ray studies

Determinations of the degree of crystallization in rubber specimens on stretching were performed with a specially arranged X-ray set up. Unstretched samples were found to be amorphous but after a deformation of about 200 per cent, crystallization set in, it increased rapidly in an almost linear fashion and then levelled off at a value of 20-25 per cent. Indications showed up that the crystal pattern differed between the A- and β-polymers. An apparatus for melting point determinations was also constructed. Preliminary tests gave a higher melting point for the A-polymers as compared with the β-polymers (ref.6).

Swelling in organic liquids

In connection with the work on copolymerization of chloroprene with various acrylic monomers, determinations of the cohesive energy density of the vulcanized copolymers were made. Swelling measurements in aliphatic liquids gave for Neoprene E, 270 J cm^{-3}; Perbunan, 315 J cm^{-3}; and for a copolymer containing 16 per cent acrylonitrile, 310 J cm^{-3}. Thus the copolymer was found to be very similar to Perbunan in its behaviour with aliphatic compounds. Swelling in alicyclic and aromatic liquids gave for an A-polymer the value 380 J cm^{-3}, thus considerably higher than the above value for Neoprene (ref.6).

Plotnikow effect

The Russian scientist Plotnikow claimed that he had discovered a longitudinal scattering when infrared light passed through a solution of threadlike macromolecules or a thin sheet of the same material. The third power of the area of the circular cross-section of the divergently scattered light was a measure of the molar mass of the macromolecules. This proposition was tested very carefully on several systems. Any scattering observed could be interpreted as a Tyndall effect (ref.6).

Summing-up

Many of the results obtained from the experimental techniques used to characterize the polymers produced as exemplified above were of a tentative nature. One significant aspect in the work was the ambition of The Svedberg to test every possible method with respect to its potential contribution to giving a most complete description of the materials studied. After the end of World War II in 1945 the circumstances changed very much both politically and economically. Therefore large parts of the work could not be continued. In the spring of 1946 The Svedberg and I had the opportunity of making a study tour of the U.S.A. It was an exhilarating experience to visit the various laboratories for polymer research, to learn about the achievements made there during the war and to compare with our own results. Also the first international meetings in the macromolecular field were of great interest. The initial gathering was held in November 1946 in Strassbourg, Colloque des Hauts Polymères.

The Svedberg was anxious to find ways of continuing the investigations on synthetic polymers. Attempts were made to organize a research institute for rubber and plastics but the efforts gave no result. Instead some of the Swedish rubber factories in 1950 formed a foundation which for a five-year period supported research on synthetic polymers. During the last years of the war some research work on poly(methyl methacrylate) had been done at the request of AB Bofors Nobelkrut. Therefore the study of this polymer became the main subject of research for some time. It was continued even after The Svedberg in 1949 retired from his chair in physical chemistry.

POLY(METHYL METHACRYLATE)

Molecular size distribution

From the work of Signer and Gross (ref.5) it became quite clear that the ultracentrifuge could be used as a tool to give information on the distribution of molecular size of thread-like molecules. When applied to two commercial samples of poly(methyl methacrylate) having sedimentation constants 37 and 58 s in acetone and molar masses 570,000 and 1,170,000 the concentration gradients showed a very irregular pattern (Kinell, refs.13, 14). The gradients covered the whole depth of the cell and indicated the existence of two or three maxima in the size distribution. This was confirmed by separating the samples in a number of fractions. Both from the sedimentation constants and the viscosity numbers of the fractions, frequency curves were obtained showing at least three peaks. I remember that these results aroused some critical interest from the research people at Imperial Chemical Industries Ltd in U.K. because such behaviour would not be expected from current theories of the kinetics of the polymerization process.

Later on Eriksson (refs.15, 21) performed some bulk polymerizations of methyl methacrylate to various degrees of conversion. The samples were separated into fractions. The frequency curve (on a viscosity number basis) showed one skew peak at conversions less than 30 per cent. At about 50 per cent a new peak appeared at high viscosity numbers. This became more pronounced at 80 per cent. After the polymerization had ceased at 94 per cent, three rather flat peaks remained. On the other hand redox polymerization in an emulsion gave polymers with a mass distribution having only one maximum (Eriksson, refs.17, 18). Then the conclusion was drawn that the appearance of several maxima was connected with the Norrish Smith-effect (ref.16).

Eriksson (refs.17, 18) also compared the experimental mass distributions with those calculated from the kinetics of the polymerization process. Furthermore he improved the method introduced by Signer and Gross (ref.5) to analyse the sedimentation diagrams for molecular mass distribution by taking the effect of the hydrostatic pressure in the cell into account (ref.19). The existence of distributions with several maxima was also shown by Claesson (ref.22) from adsorption analysis of poly(methyl methacrylate).

Shape and structure of the macromolecules

In connection with ultracentrifugation it was of interest to study the dependence of the sedimentation constant on the molar mass of the molecules. Kinell (ref.13) found for the two commercial poly(methyl methacrylate) samples a rather irregular relation between the sedimentation constant and the viscosity number, which could not be analysed in a rational way. A more extensive study was made by Eriksson (refs.20, 21). He found a linear log s/log[η] relation with a slope of 0.78 up to a certain molar mass. At higher viscosity values a non-linear deviation was observed. He attributed this to a change in configuration of the macromolecules formed at the explosive stage of the polymerization.

Ultraviolet absorption measurements in the spectral region 250-350 nm on poly(methyl methacrylate) samples prepared through initiation by benzoyl peroxide (Kinell, refs.23, 24) showed that the end-groups were benzoyloxy and phenyl radicals. The measured absorptivities at 273 nm and molar masses from viscosity data gave the number of benzoyloxy end-groups. Samples prepared at 80-90°C gave values between 0.6 and 0.8. Some infrared and raman measurements were also performed (ref.23).

MISCELLANIES

In parallel with the investigations reviewed above some topics of a more general nature were treated. Thus the dB/dx-method introduced by Gralén (ref.25) to characterize poly-molecularity from sedimentation measurements was improved (Eriksson, ref.18; Kinell refs.14, 26) with respect to various influences during the course of sedimentation. Furthermore some problems concerning the effect of concentration dependence and hydrostatic pressure on the radial dilution in the ultracentrifugal cell have been treated (Eriksson, ref.19; Kinell refs.27, 28). Finally a contribution has been given to the discussion of the types of molar mass averages which can be obtained from sedimentation velocity and diffusion measurements on threadlike molecules (Kinell, ref.29).

SOME PERSONAL REFLECTIONS

After this short review of the work that was performed on synthetic polymers in The Svedberg's laboratory I would like to add some personal reminiscenses from the time when I was his assistant. I joined his group working on the polychloroprene project in 1943. Before that I had never met him. I had only from time to time observed him come and go between the institute and his residence. Also I attended a series of lectures that he gave in 1935 on the chemistry of the atomic nucleus after the nuclear transformations had been discovered. My first task became to penetrate the current theories of rubber elasticity and to plan for some experiments on the elastic behaviour of polychloroprenes. Later on I took over part of Professor Brohult's responsibilities in coordinating the research work.

It was very stimulating to have The Svedberg as a superior because he inspired everyone with great confidence. He very much appreciated being informed about the progress of the work and greeted every new idea and suggestion enthusiastically. Sometimes he appeared shy. I remember one occasion at a time when the research activities were at a maximum. Work was going on not only in physical chemistry but also in biochemistry and in medical chemistry. The institute was crowded with people, several persons shared the same laboratory room and also the corridors were made use of. One Sunday when I was doing some work he entered my room and asked in a very humble way if I could guide him through the institute and tell him which person was working in what room and on what subject. He said he felt very embarrassed at not having a clear view of the situation at the institute. I had the responsibility of allocating space and facilities to all personnel and therefore, he thought, I could help him out of his quandary. Really it became something to remember spending that Sunday afternoon with him as a disciple of a most inquiring mind.

Sometimes The Svedberg would get furious and angry about things especially when the reason for an occurrence was found to be due to carelessness or neglectfulness. Mostly, however, he showed a modest and very kind attitude towards those closest to him and even towards persons that he met occasionally. In the summer of 1945 I accompanied him to observe the total eclipse of the sun at Skelleftehamn in northern Sweden. At the same time he had planned some botanical excursions and one day we went by car to Lycksele in Lappland. We spent the night at a small hotel. Next morning we were informed that breakfast could not be served because during the night a thunderstorm had caused some damage to the kitchen. Instead we were recommended a small eating-place within walking distance of the hotel. When we arrived there the landlady greeted us and turned to Svedberg whispering that she was very proud this morning because one of her guests at the next table was the forest officer of the Mo och Domsjö Co. Svedberg answered that he felt very happy on her behalf for this event. In the afternoon when we came back from our excursion Svedberg suggested that we have our dinner at the small eating-place. Now the landlady greeted us with great enthusiasm and we had not even taken our seats before she went from one table to another whispering to her other guests and nodding at our table. We were in no doubt that the landlady experienced one of the most exciting days of her life and quite obviously Svedberg felt very happy about this.

From a short review by The Svedberg in 1946, ´Forty Years of Colloid Chemistry´ (ref. 30), I quote: "The growing importance of synthetic high molecular weight substances as the material for an ever increasing number of technical products has accentuated the need for information. One of the most ardently pursued lines of research in present-day colloid chemistry is therefore the elucidation of the rules governing the formation of organic macromolecules of different build and the determination of their structure". Knowing today, almost forty years after this was written, about the tremendous evolution within the field of synthetic polymers one may ask if we – the members of Svedberg´s group for research on synthetic polymers – really were devoted enough to our tasks, if we could completely understand his foresight, if we could follow him in his eagerness to find new methods and techniques. It is difficult to find an answer even in self-examination. Svedberg seldomly criticized, he encouraged new efforts and thoughts. I have a feeling that I made him disappointed at one occasion. When the first communications on nuclear magnetic resonance phenomena appeared he asked me to give my opinion about the application of this technique to our work. I read the reports and answered him that I was in doubt about its possible use to us mainly because of the amount of advanced physics that was involved. He accepted my views and did not return to the subject until 1952 when Bloch and Purcell got the Nobel prize in physics for their discoveries. I have never stopped wondering what intuition Svedberg had about the potential of the method. Today we know that nuclear magnetic resonance is one of the most accurate methods of obtaining information about polymer structure.

I am grateful to have had the opportunity to begin my research career with The Svedberg as a teacher – it was a rich time both from a scientific and a human point of view.

Acknowledgement – In this short survey it has not been possible to do full justice to the contributions made by all the people involved in the synthetic rubber programme. However, all of them are worthy of every recognition for their most unselfish and devoted engagement in the work.

REFERENCES

1. T. Svedberg, Meddelande från Sveriges Kemiska Industrikontor **25**, 126-129 (1942).
2. H. Staudinger, Die hochmolekularen organischen Verbindungen – Kautschuk und Cellulose, p. 101, Springer, Berlin (1960).
3. R. Signer and R. Gross, Helv. Chim. Acta, **17**, 59-71 (1934).
4. R. Signer and R. Gross, Helv. Chim. Acta, **17**, 335 (1934).
5. R. Signer and R. Gross, Helv. Chim. Acta, **17**, 726-735 (1934).
6. T. Svedberg and P.-O. Kinell, Undersökningar över syntetisk kautschuk, Uppsala (1946), (Typed report in Swedish).
7. T. Svedberg and P.-O. Kinell, Harald Nordenson 60 år, p. 321, Esselte, Stockholm (1946).
8. P.-O. Kinell, J. Phys. Colloid Chem., **51** 70-79 (1947).
9. K. Granath, Acta Chem. Scand., **4** 103-125 (1950).
10. S. Ekegren, O. Öhrn, K. Granath and P.-O. Kinell, Acta Chem. Scand., **4**, 126-139 (1950).
11. B. Vollmert, Grundriss der makromolekularen Chemie, p. 247, Springer, Berlin (1962).
12. L.R.G. Treloar, Trans. Far. Soc., **36**, 538 (1940).
13. P.-O. Kinell, Acta Chem. Scand., **1**, 832-847 (1947).
14. H. Mark and E.J.W. Verwey, Adv. Colloid Sci. III, p. 161, Interscience, New York (1950).
15. A.F.V. Eriksson, Acta Chem. Scand., **3**, 1-12 (1949).
16. R.G.W. Norrish and R.R. Smith, Nature, **150**, 366 (1942).
17. A.F.V. Eriksson, Acta Chem. Scand., **7**, 377-397 (1953).
18. A.F.V. Eriksson, Acta Chem. Scand., **7**, 623-642 (1953).
19. A.F.V. Eriksson, Acta Chem. Scand., **10**, 360-377 (1956).
20. A.F.V. Eriksson, Acta Chem. Scand., **10**, 378-392 (1956).
21. A.F.V. Eriksson, Svensk Kem. Tidskr., **68**, 1-22 (1956).
22. S. Claesson, Arkiv Kemi, **A26**, No 24 (1949).
23. P.-O. Kinell, A spectrophotometric study of poly(methyl methacrylate), Almqvist & Wiksell, Uppsala (1953).
24. P.-O. Kinell, Arkiv Kemi, **14**, 353-370 (1959).
25. N. Gralén, Sedimentation and diffusion measurements on cellulose and cellulose derivatives, Almqvist & Wiksell, Uppsala (1944).
26. P.-O. Kinell, Acta Chem. Scand., **1**, 335-350 (1947).
27. P.-O. Kinell, J. chim. phys., **44**, 53-57 (1947).
28. P.-O. Kinell, Arkiv Kemi, **14**, 295-304 (1959).
29. P.-O. Kinell, Arkiv Kemi, **14**, 327-336 (1959).
30. T. Svedberg, Harald Nordenson 60 år, p. 343, Esselte, Stockholm (1946).

APPENDIX

As a complement to the presentation in this paper of Svedberg´s research work on synthetic polymers I would like to add a few comments on the technical development which lead to production of the synthetic rubber Svedoprene[R]. The rubber project involved mainly three steps. After the first exploratory work on the laboratory scale to find the feasibility of synthesizing monovinylacetylene, chloroprene and polychloroprene, a small pilot plant was erected in one of the corridors of the Institute of Physical Chemistry. Based on the experience gained on the various possible processes as well as the best construction materials to be used, the Swedish Rubber Research Laboratory at Uppsala was set up in an old laboratory building (Head: Göran Philipsson, M. Eng.). At this site, facilities were created to develop the production processes on a still larger scale (In Charge: Bengt Naucler, M. Eng. and Tor Sundqvist, M. Eng.). Some production of Svedoprene[R] rubber was started there. Parallel with this pilot plant work, studies of catalyst systems for monomer production were initiated at the LKB Laboratory in Stockholm (Head: Sven Brohult, Professor). Testing of polymerizates produced by the pilot plant was performed by the rubber factories at Helsingborg and Trelleborg in Southern Sweden.

When sufficient experience had been gained from the pilot plant work, plans were worked out for a full scale commercial plant by the Stockholm Superfosfat Fabriks Aktiebolag works at Ljungaverk (Gustav Carlsson, Managing Director) in the province of Medelpad in Northern Sweden. This company already produced calcium carbide from which acetylene was obtained. The rubber plant began operating in midsummer 1944. Later on a second plant was erected by the same company at Stockvik, close to Sundsvall. Production of Svedoprene[R] was discontinued in 1945 when the war restrictions on free trade were raised.

P.O.K.

8. The First Svedberg Centrifuges

J. Burton Nichols

1505 River Road,
Wilmington, Delaware, USA

Early in 1923 The Svedberg was invited to the University of Wisconsin to conduct six months of research and give a series of lectures in the young field of colloid chemistry. On his way out he stopped at the Eastman Kodak Co. in Rochester, N.Y. to give a general lecture on his work in Sweden.

At the time I was just finishing my undergraduate work at Cornell University with Wilder D. Bancroft as my major professor, so I went to Rochester to meet Svedberg. I expected to see a plump, middle-aged bewhiskered European professor with grey hair. Imagine my surprise to meet this boyish, slight man with flashing eyes and very black hair (he was 38 at the time). In about two weeks I was at Wisconsin and had selected the optical centrifuge as my research project. Svedberg told me later that the other men had avoided it for more promising projects. But it seemed just right to me.

The list of projects he brought with him represented his current interests. These included formation of colloidal metals by high frequency arc and the formation of Mn arsenate gels under the kinoultramicroscope, which was investigated by Elmer O. Kraemer, later in charge of physical colloid research at the DuPont Co.; modification of the sedimentation tube to examine emulsions, investigated by Alfred J. Stamm, later an authority on the physical structure of wood; and characterization of colloidal clays, investigated by Richard Bradfield, later the renowned soil scientist at Cornell. Jack Williams worked on specific heats or organic liquids and later became a Wisconsin professor with a very active ultracentrifuge research group starting in the late 1930s.

The lectures were very well attended by academic and industrial people from all over the country. I can still see The Svedberg touching a 500 V DC bare wire to see if it was live for a lecture experiment. The lectures appeared in book form the next year.

In June Prof. J. Howard Matthews, head of the Chemistry Department, brought together colloid chemists from throughout the USA and Canada for the first of a long series of annual colloid symposia, with Svedberg as the honoured guest professor. It was very successful. In the Spring and Summer he took a little time for picnics, exploring the countryside and photographing the Wisconsin wild flowers.

By the time he returned to Sweden in August the crude little optical centrifuge had been developed to a stage where pictures could be taken of sedimenting gold sols and clays but convection was a serious problem. The centrifuge's chief value lay in the orientation it gave to the problem. The heavy cylindrical rotor was directly connected to a vertically mounted 20,000 r.p.m. Dumore motor yielding only about 2,000 r.p.m. because of the rotor weight. The two arms were slotted vertically and horizontally so that direct or scattered light could be observed. The photo- and cross-section give an idea of the construction (Fig. 1, 2 and 3). The 3-inch test tubes used for the sols led to convection as well as sedimentation (Fig. 4).

Svedberg took back with him the problems to be solved: vibration, convection, temperature control, and cell construction, but with confidence that a machine could be built to give reliable data on the size distribution of very fine-grained colloids. His eight-month stay at Wisconsin had been very stimulating, not only in ideas for a functioning centrifuge worked out on board a ship to Sweden, but also for electrophoresis, diffusion and other phenomena in colloid chemistry.

In the intervening two years until I came to Uppsala in 1925, he and Herman Rinde developed the first low speed ultracentrifuge with sector-shaped cells, vibrationless mountings of a modified cream separator with self-balancing rotor, and hydrogen atmosphere in the rotor chamber to equalize the temperature (Fig. 5 and 6). Convectionless sedimentation of Faraday gold sols was achieved!

In 1924 Svedberg and Robin Fåhraeus, Professor of Pathology at Uppsala University, were able to establish a sedimentation equilibrium of CO-haemoglobin from which they calculated a molecular weight of 68,000 or 4 x 17,000, the value obtained from the iron content. This value was substantiated by Adair in England by careful osmotic pressure measurements. This was the first indication that the soluble proteins are of uniform molecular size and probably structurally uniform giant molecules. This finding became one of the bases of all the succeeding work leading to the present field of molecular biology.

My first assignment in Uppsala was egg albumin. Since it is a colourless protein it was necessary to use ultraviolet light corresponding to an absorption band in the protein spectrum (Fig. 7). Again it appeared to be a uniform giant molecule (later verified by the more sensitive scale-method to be about 47,000).

Fig. 1.   Svedberg´s and Nichols´
optical centrifuge (Madison 1923).

Fig. 2.   Dr. J.B. Nichols, Rockefeller fellow
in Uppsala, works with the ultracentrifuge
in a toilet converted to laboratory
(Dagens Nyheter, Oct. 1926).

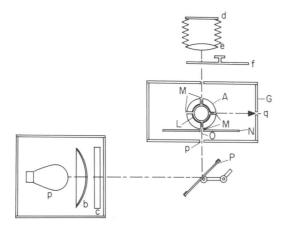

Fig. 3.   Light path of the optical
centrifuge (Madison 1923).

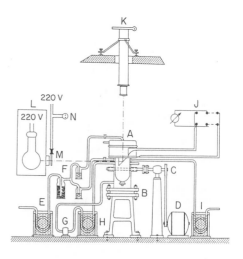

Fig. 4.   Light beam through gold sol
in the test tube of the optical centrifuge
(Madison 1923).

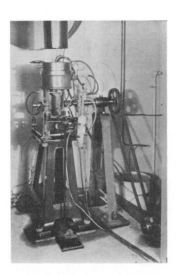

Fig. 5 and 6.   The first De Laval optical
ultracentrifuge built in Uppsala
(T. Svedberg and H. Rinde 1925).

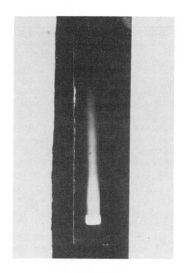

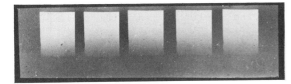

Fig. 7

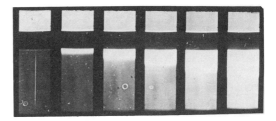

Fig. 7 and 8. Sedimentation of egg albumin (top) and haemocyanin from the vineyard snail (right) in the ultracentrifuge (T. Svedberg, J.B. Nichols and E. Chirnoaga, 1926).

Fig. 8

A surprising result was obtained by Chirnoaga on haemocyanine from vineyard snails. It sedimented with a sharp boundary (Fig. 8) indicating a molecular weight of nearly 7,000,000 instead of a value of about 16,000 based on its copper content. (At one time after we had pressed out the blood of the snails, we had a feast of baked snails au gratin and Chianti wine.)

To obtain better resolution and shorten the time from days to hours, Svedberg developed the Oil Turbine Ultracentrifuge in 1926 with the help of engineer A. Lysholm of the Ljungström Steam Turbine Co. and instrument maker Ivar Eriksson's skillful machining (Fig. 9, 10 and 11). It was designed for 40,000 r.p.m., but the first tests gave only 19,000. After a long series of changes in rotor turbine and bearing design, he reached the goal of 40,000 r.p.m. yielding a centrifugal force close to 100,000 times gravity. A determination of molecular weight of CO-haemoglobin from the sedimentation velocity and diffusion constant could then be made in three hours (Fig. 12).

Fig. 9.

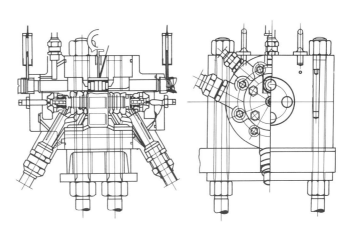

Fig. 10

Fig. 11.

Fig. 9, 10 and 11. Oil turbine ultra-centrifuge: picture, cross section and rotor (T. Svedberg, 1926).

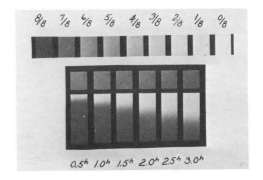

Fig. 12. Sedimenting CO-haemoglobin in the oil turbine ultracentrifuge with concentration scale on top.

In 1926 Svedberg and his group of Swedish and foreign associates had this picture taken
in front of Nya Kemikum (Fig. 13). Left to right: Arne Tiselius, Herman Rinde, Neil B.
Lewis (Australia), Tominosuke Katsurai (Japan), E. Chirnoaga (Roumania), Olle Quensel, The
Svedberg, J.B. Nichols (USA), and Hugo Anderson (Fig. 13). Anderson photographed the Códex
Argenteus, page by page, to preserve a record of this 450 A.D. illuminated book of the four
Gospels in Meso Gothic.

The same group was photographed on the hill called "Slottsbacken" in Uppsala in June at
12:20 a.m. (20 min. after midnight) after an evening at a cafe. Svedberg received many
visiting scientists at that time (Fig. 14).

Arne Tiselius, as a new doctoral candidate, acted as unofficial receptionist for us as
we came, and met Katsurai at the railroad station in response to a call from Stockholm.
Arne asked him how long he intended to stay because the little Katsurai came with only
several books tied with a silk scarf. Katsurai responded, "Two years, if I may". Arne was
somewhat taken aback.

In October 1926 an illustration of Svedberg´s talents in raising money for the
institute appeared on the front page of Dagens Nyheter, the Stockholm newspaper. The
article was headed: "Professorn får arbeta i mörk kolkällare och hans lärjungar på W.C."
(The professor has to work in a dark coal-cellar and his students in a toilet.) The next
month he was awarded the Nobel Prize in Chemistry and in December, the Swedish Riksdag (the
parliament) found funds for a new Institute of Physical Chemistry with adequate and well-
arranged space for the burgeoning ultracentrifuge studies and the other research projects in
progress (Fig. 2).

After Svedberg was awarded the Nobel Prize in late 1926, he considered changing his
means of transport from a bicycle to a car. When he passed his driving test, about ten of
us took him to dinner at Flustret to celebrate his success. He was frequently toasted. On
our way back to Kemicum we decided to serenade a popular Göteborg coed, Gertrud Mannheimer
from Gothenburg, Sweden. As we sang we noticed that The was becoming more and more
agitated. It turned out that we were serenading without a permit and might be arrested if a
policeman came along and heard us. No such luck!

I hope this sampling of the early years with The Svedberg at Wisconsin and then at
Uppsala will give you some of the flavour of the times - not all work, some play.

Fig. 13   Fig. 14

Fig. 13 and 14. T. Svedberg´s research group in 1926, in front of the new
chemistry building (Uppsala University) and T. Svedberg´s picture of his group
the same year (at midnight, June 1926, long exp.).

9. Contributions of Svedberg to Nuclear and Radiation Research

B. Larsson and S. Kullander

The Gustaf Werner Institute, Uppsala University, Uppsala, Sweden

Abstract – With the desire to understand the submicroscopic and molecular nature of matter, The Svedberg extended his scientific curiosity and endeavour into the domain of nuclear and radiation sciences. Among many personal contributions, only a few of the more important are reviewed in this paper, namely the anticipation of the isotope concept, a theory for the action of light on the photographic emulsion, and a model describing the action of ionizing radiation on macromolecules.

Also indirectly, especially in the years following the Second World War, Svedberg continued to contribute to the advance of nuclear chemistry and physics. He created the Gustaf Werner Institute, where the 185 MeV synchrocyclotron was built. Under his leadership several important discoveries and technical or medical developments were then made. The present scientific programme for the new cyclotron laboratory and related activities at the European Centre for Nuclear Research (CERN) continue to add evidence of his scientific foresight.

As the memory of The Svedberg is so closely associated with pioneering in colloid research, and the invention of the ultracentrifuge, his contributions to other fields of science and technology may be unduly neglected. In this lecture, attention is focused on some significant results and initiatives in nuclear and radiation sciences, where Svedberg´s intuition, experimental skill and vision are of enduring importance.

When Svedberg matriculated at Uppsala University at the age of 19, early in 1904, nuclear science was still in its childhood. Only eight years previously, Becquerel had made the revolutionary observation that radiation from uranium causes fogging of a photographic plate. The electron was known in 1897, thanks to Thomson, but the Rutherford–Bohr model of the atom was not to be proposed until 1913, the year when Svedberg as professor of physical chemistry in Uppsala gave his inaugural lecture, entitled "Atomic Research in Modern Physics and Chemistry"!

RADIOACTIVE ELEMENTS AND THE ISOTOPE CONCEPT

In ten years of inspiring experimental research before the First World War, Svedberg was occupied with phenomena believed to reflect the molecular structure of matter. This is well demonstrated in the publications of that time (ref.1). The underlying strategic idea was that chemical phenomena might be understood from first principles, if the microscopic aspects of the properties of matter were to be clarified in precise physical terms.

Little wonder that Svedberg felt challenged when Rutherford and Soddy revealed, in 1902, that an essential feature of radioactivity is the spontaneous transformation of one chemical element into another. This event showed the significance of the discovery of radioactive elements by Marie and Pierre Curie. Their endeavour, at the dawn of the "nuclear age", stimulated an interest in radioactivity which Svedberg carried with him throughout his life.

In an attempt to fit the intermediary products of the radioactive disintegration series to the periodic system, Svedberg set out, with D. Strömholm, to find chemical correlations between them and the known stable elements (ref.2). In experiments reported in 1909, they found that Thorium X and Actinium X crystallized with lead and barium salts but not with other salts tested (Table 1). Thereby they were probably the first to indicate the existence of atoms with different physical properties in the same chemical element. It was left to Soddy to introduce the isotope concept on the basis of his own more conclusive investigations. In fact, Soddy generously stated in his Nobel lecture, in 1922:"... Strömholm and Svedberg were the first to suggest a general complexity of the chemical elements concealed under their chemical identity. Until I read their paper again, in the preparation of this lecture, I had not realized how explicit this anticipation of present views is". Strömholm and Svedberg probably had regrets that the Nobel committee had not realized this fact either!

Table 1. Svedberg and Strömholm had already anticipated the isotope
concept in 1909, by indicating that three different "radioelements" show
chemically identical behaviour. "Thorium X", "Actinium X" and "Radium"
were shown to coprecipitate from a concentrated solution of barium
nitrate (From ref.2b).

Versuche mit Ba(NO$_3$)$_2$ (je 2 Proben).

Element	Aktivität		Verhältnis
	der Krystalle	der Mutterlauge	der Aktivitäten
ThX	45.3	30.0	1.5
	13.0	8.2	1.6
Akt. X	134.1	87.6	1.5
	39.6	24.5	1.6
Ra	150.0	102.0	1.5
	19.4	12.1	1.6

RADIATION RESEARCH

In the early years of his career, Svedberg also had experience in the use of radiation as a
tool for the experimental modification of chemical systems, and for analytical purposes. The
first study of this kind (ref.3) represents only a modest facet in Svedberg´s investigation
of metal colloids. The hope was that the exposure of suspensions of metal powder to UV-
light or X-rays could be a means of facilitating the production of colloidal solutions. The
evidence found was not conclusive and the ideas did not find any applications. However, one
important indirect result - not at all trivial - emanated from these experiments: a
profound experience in the use of radiation in laboratory work. Svedberg´s familiarity with
radiological matters could not find a more beautiful illustration than in publications
written with the photographer I. Nordlund (ref.4). Here it was shown that old deteriorated
printing could be made legible by utilization of the contrast obtained by means of UV-
fluorescence or X-ray absorption. The findings were successfully applied to the interpretation of the severely impaired text of the Silver Bible, Codex argenteus, in the Uppsala
University library (Fig. 1).

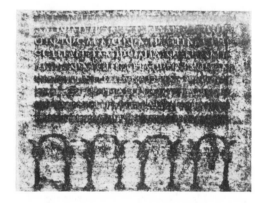

Fig. 1. From the photographic study of
the Silver Bible. The original (above)
has been made legible (below) by the use
of ultraviolet light and fluorescence
(ref.4).

In the years after the First World War, Svedberg gained professional ability in photography, not unusual among scientists at that time. This experience formed the basis of his profound interest in the photographic process, in particular the action of light on the photographic emulsion. By microradiography with light and X-rays of partially developed silver bromide grains he could demonstrate multiple active centres in the individual halide crystals (Fig. 2). On the basis of such observations Svedberg and H. Andersson worked out the well-known theory for the latent image in the primary photographic process (ref.5).

In 1938 Svedberg and his student S. Brohult studied the splitting of the large protein molecule haemocyanin with UV- and alpha-radiation, respectively (ref.6). They had hoped to relate the observed splitting of weak bonds between protein sub structures to the absorption of single UV-radiation quanta. This failed, but for the splitting of haemocyanin by alpha rays, a simple single-hit model could be easily applied. The theoretical model described may be considered as a "target hypothesis". This later came to play an important role in the interpretation of experimental results in radiation chemistry and biology, in terms of numbers of "hits" on geometrically defined targets in the irradiated structures considered.

Svedberg naturally foresaw, on the basis of his own experience, that subatomic particles from accelerators would become indispensable tools in radiation physics and chemistry. Available in high currents and at variable energies they would soon become superior to the alpha particles from polonium that Svedberg and Brohult had used to irradiate haemocyanin. A special challenge was of course the observation in 1932, by Chadwick, that neutral particles of nuclear origin, neutrons, could be produced by bombarding beryllium with alpha rays. Svedberg was 50 years old when Frédéric and Irène Joliot discovered artificial radioactivity, four years after E. Lawrence's invention of the cyclotron, and two years after the first nuclear transmutation with the aid of a particle accelerator. The period before the Second World War is still referred to as the "golden age" of nuclear science.

What could be more natural to Svedberg than the idea of building a small particle accelerator in Uppsala? The task was given to one of his students, H. Tyrén, later successor to Svedberg as director of the Gustaf Werner Institute. The year was 1941 when it was decided to build a neutron generator based on the bombardment of deuterated ice with 200 keV deuterons. The construction was completed in 1944 (ref.7). In this year, Uppsala thus got its first ion accelerator (Fig. 3). This apparatus was used for irradiation purposes and for the production of some radionuclides for tracer work but it was bound to become obsolete in a short time.

At this time the 60-year old professor of physical chemistry was contemplating what to do after his mandatory retirement at the age of 65. A fine solution to this problem occurred when his friend, the professor of obstretics and gynaecology J. Naeslund,

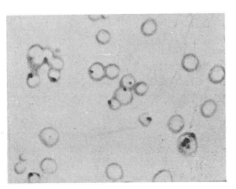

a.

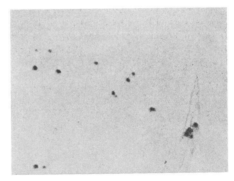

b.

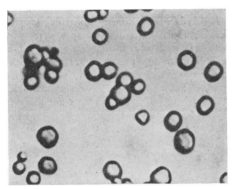

c.

Fig. 2. From a microscopic study of metallic silver centres appearing during the development of exposed grains of silver bromide (ref.5). Consecutive photographs of the halide crystals (a) and silver grains (b) were superimposed to indicate the position of the active centres (c).

suggested that they should approach a well-known industrialist in Gothenburg, Gustaf Werner, with an application for financial assistance to build a large cyclotron in Uppsala. Svedberg´s assistant Tyrén who was then studying in New York City was asked to go to California and collect essential details of design at Berkeley, already a centre of accelerator technology.

In December 1949, soon after Svedberg´s 65th birthday, a 185 MeV proton synchro-cyclotron (Fig. 4) was inaugurated by the Swedish Crown Prince, later King Gustaf VI Adolf. Uppsala had got two unique assets for nuclear and radiation research: a young professor emeritus, boiling with enthusiasm over the potentialities of nuclear and radiation research, appointed director of the Gustaf Werner Institute for unlimited time, and a most powerful accelerator!

Fig. 3. Assembling of the first particle accelerator in Uppsala, later used for the production of neutrons with 200 keV deuterons (ref.7). The Svedberg (left) is seen here with his collaborators Lars Mattsson, Helge Tyrén and Sven Brohult.

Fig. 4. The synchrocyclotron at the Gustaf Werner Institute, completed in 1951 (ref.8).

THE GUSTAF WERNER INSTITUTE

Svedberg was thus the founder of the Gustaf Werner Institute, which rapidly developed into a
radiation research centre. The first beam of 185 MeV protons from the synchrocyclotron
(ref.8) was obtained in 1951 and for a while Uppsala had the highest particle energies in
Western Europe, surpassing the facilities in Amsterdam and Harwell, England by 50 and 170
MeV, respectively.

   Among the scientific highlights in the history of the Institute was the first direct
observation in 1956 of proton energy shells in nuclei. These were found in detailed
measurements of proton knock-out reactions from light nuclei (ref.9). Tyrén and his
collaborators had for the first time measured not only the angles but also the energies of
the two emerging protons (Fig. 5).

   In 1957 biological experiments and medical therapy using 185 MeV protons were introduced by B. Larsson and collaborators (ref.10). Protons of this energy have a range of 24 cm
in tissue and are therefore well suited to medical application. Radiation surgery with
pencil-shaped proton beams (ref.11) was successfully carried out on patients suffering from
Parkinson's disease and other brain disorders (Fig. 6). Infiltrating malignant tumours
were treated with large fields of range-modulated protons (ref.12).

   In the early 70's it was found that pions were produced in proton nucleus collisions
below the threshold for such production in free nucleon-nucleon collisions, leading to discrete
nuclear states (ref.13). Since then this type of nuclear reaction has been studied in many
laboratories.

   The Gustaf Werner Institute also participated from the very start in the 50's in
the work at the joint European high-energy physics centre, CERN, in Geneva.

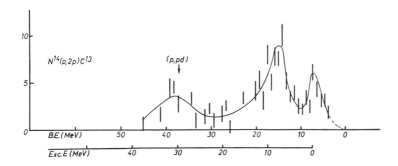

Fig. 5. From a study of the reaction of 185 MeV protons with atomic
nuclei of nitrogen-14. The varying binding energy (B.E.) of a removed
proton is indicated as well as the excitation energy (Exc. E.) of the
residual nucleus (ref.9).

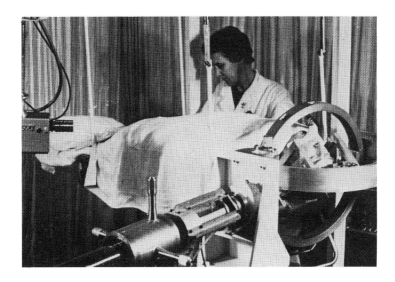

Fig. 6. Use of 185 MeV protons for cerebral surgery (ref.11).

The Gustaf Werner Institute has had a most active involvement with CERN. The first accelerator was built by accelerator technicians and physicists from the Gustaf Werner Institute, in collaboration with specialists from the Philips company and the accelerator research laboratory at Harwell, England. When the CERN synchrocyclotron was reconstructed in the middle of the 60´s, engineers from Uppsala also participated. From this time, Uppsala physicists have been continuously involved in various experiments at the CERN accelerators (Fig. 7).

The synchrocyclotron was shut down in 1977 for complete reconstruction. The original flat-pole geometry has now been changed to a three-sector spiral ridge geometry to make possible a radially increasing magnetic field while maintaining axial focussing. With this new field configuration, protons starting at a frequency of 24 MHz reach a final kinetic energy of 200 MeV at a frequency of 21.5 MHz. In the earlier flat-pole geometry, the frequency range 33-26 MHz was achieved using a rotating condensor, resulting in an extracted beam of 185 MeV. The new accelerator (ref.14) will have broad-band amplifiers for the acceleration of protons. The frequency modulation is done on a low signal level before amplification. Other particles will be accelerated in the cyclotron-mode with constant frequency by adjusting the magnetic field with correction coils. Intensities of 10 to 100 µA are expected depending on the particle accelerated and on the energy chosen. The new beam properties will be far superior to those of the old accelerator which gave only protons of a fixed energy and extracted beam intensities of 10 to 100 nA.

Presently the programme of the Institute covers many different disciplines of natural science and medicine, such as high energy physics, intermediate energy physics, nuclear chemistry, physical biology, oncology and neurosurgery. Clinical research programmes are pursued with the collaboration of physicians from Uppsala, Stockholm, Harvard and Moscow.

In the new experimental areas, 1800 m^2, adjacent to the synchrocyclotron (Fig. 8), experimental equipment is now being installed. Physicists from the Institute are preparing an experimental set-up for the study of "Bremsstrahlung" by protons in the strong nuclear field and a high-quality beam for neutron-induced reaction studies, to mention two of the most advanced projects in physics. In the same time new biological and medical applications of accelerated particles and radionuclides are being prepared. Target areas have been designed for work on living cells, animal tumour models and patients from the departments of oncology and neurosurgery.

Nearby an additional hall has been built in which the CELSIUS storage ring will be placed. CELSIUS in an acronym for Cooling with Electrons and Storing Ions from the Uppsala Synchrocyclotron. The basic structure of the CELSIUS ring was taken over from CERN, in 1983, where it was used to demonstrate that antiprotons can be cooled. The CELSIUS ring is expected to operate in three years´ time. It will bring in a new method in the field of nuclear science, namely the use of a very intense circulating internal beam colliding with ultra-thin targets crossing the circulating beam. Laboratories in Bloomington, Indiana and Tokyo have also started similar projects and many laboratories are contemplating this new technology.

Fig. 7. The first accelerator at CERN, a 600 MeV synchrocyclotron, was constructed in collaboration with engineers from the Gustaf Werner Institute. Now the international nuclear research centre in Geneva has grown into an outstanding facility with several large accelerators as indicated on this aerial view.

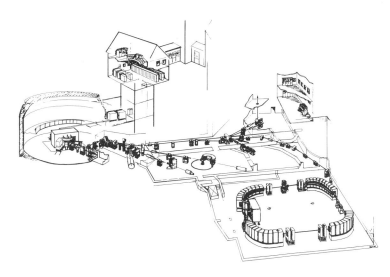

Fig. 8. The new cyclotron laboratory at the Gustaf Werner Institute
with its various beam-lines for physical, chemical and biomedical
research. A magnet ring CELSIUS is indicated in the lower right
corner of the drawing.

Some examples of the present research programme show diverse activities. Scientists
from the Institute participate in two important projects involving electron-positron
collisions. In one of them, they participate at CERN in the construction of a big detector,
DELPHI, for studies of electron-positron collisions at 100 GeV. Such collision energies,
higher than ever, will be obtained in 1989 in the 27 km underground ring called LEP, Large
Electron-Positron, colliding machine at CERN. DELPHI will be used mainly for the study of
quarks and leptons and the basic forces acting between these most fundamental particles of
today. Maybe the results from LEP will give some clues as to whether these fundamental
elementary particles have some common sub structure.

The other electron-positron project presently in progress involves positrons of very
low energies. It makes use of the annihilation reaction between a positron from the decay
of radioactive carbon-11 nucleus and an electron in the tissue material of the body. The
two resulting gammas are detected using a so-called positron camera. This research is a
joint undertaking between the Institute's biologists and nuclear chemists, the Institute of
Chemistry and the University Hospital. The Institute is here involved in the production of
nuclides, their separation, incorporation in organic molecules and physiological use.

Physicists from the Institute participate within the European Muon Collaboration
at CERN in the study of quark and gluon structure of atomic nuclei by irradiating nuclei
with high energy muons. By studying the effects when these nucleon constituents are hit by
the muons, new information is evolving about these fundamental structures.

A last example on the present research are experiments in progress at LEAR, the Low
Energy Antiproton Ring at CERN. The antiprotons are used for the production of heavy
hypernuclei as well as for studies of interactions between hyperons and antihyperons.

CONCLUSIONS

The Svedberg's own scientific work has contributed in several ways to the 20th century
revolution in the physical, chemical, and biological sciences. Equally important, though
indirect, is the constructive and visionary scientific leadership that has led to many
important developments, mainly through the work of his collaborators at the Gustaf Werner
Institute and at CERN. The converted synchrocyclotron will continue to produce evidence of
the scientific foresightedness of the great scholar and technologist, The Svedberg, far into
the next century.

REFERENCES

1. S. Claesson and K.O. Pedersen, <u>Biographical Memoirs of Fellows of the Royal Society</u>, **18**,
    595-627 (1972).

2a.   T. Svedberg and D. Strömholm, Untersuchungen über die Chemie der radioaktiven Grundstoffe I. Z. anorg. Chem., **61**, 338–346 (1909).

2b.   T. Svedberg and D. Strömholm, Untersuchungen über die Chemie der radioaktiven Grundstoffe II. Z. anorg. Chem., **63**, 197–206 (1909).

3.   T. Svedberg, Uber die Bildung disperser Systeme durch Bestrahlung von Metallen mit ultra-violettem Licht und Röntgenstrahlen. Z. Chemie Ind. Kolloide **6**, 129–136 (1910).

4.   T. Svedberg and I. Nordlund, Fotografisk undersökning av Codex argenteus. Uppsala Univ. Årsskr. 1 Mat. Naturv., 1–26 and 22 plates (1918).

5.   T. Svedberg and H. Andersson, On the Relation between Sensitiveness and Size of Grain in Photographic Emulsions, Photogr. J., **61**, 325–332 (1921).

6.   T. Svedberg and S. Brohult, Splitting of Protein Molecules by Ultraviolet Light and Alpha-rays, Nature, Lond. **143**, 938 (1939).

7.   H. Tyrén, A Neutron Generator for Preparative Use. In "The Svedberg 1884 30/8 1944", p. 224, Almqvist och Wiksell, Uppsala (1944).

8.   B. Hedin, Synkrocyklotronen i Uppsala, Kosmos (Stockh.), **30**, 144–154 (1952).

9.   H. Tyrén, P. Hillman and Th.A.J. Maris, High Energy (p,2p) Reactions and Proton Binding Energies. Nuclear Physics, **7**, 10–23 (1958).

10.   B. Larsson, L. Leksell, B. Rexed, B. Sourander, W. Mair and B. Andersson, The High-energy Proton Beam as a Neurosurgical Tool. Nature, **182**, 1222 (1958).

11.   B. Larsson, L. Leksell and B. Rexed, The Use of High-energy Protons for Cerebral Surgery in Man. Acta Chir. Scand., **25**, 1 (1963).

12.   S. Falkmer, B. Fors, B. Larsson, A. Lindell, J. Naeslund and S. Stenson, Pilot Study on Proton Irradiation of Human Carcinoma. Acta Radiol., **53**, 33 (1962).

13.   S. Dahlgren, B. Höistad and P. Grafström, Positive Pion Production in Nuclear Reactions Induced by 185 MeV Protons. Physics Letters, **35B**, 219 (1971).

14.   S. Holm, A. Johansson and the Cyclotron and CELSIUS Groups, The Gustaf Werner Institute and the Tandem Accelerator Laboratory. In Proceedings, Tenth International Conference of Cyclotrons and their Applications, East Lansing, Michigan, (F. Marti, Ed.) pp. 589–594, New York (1984).

III Colloid and Surface Science

10. Surface Enhanced Raman Scattering by Colloidal Silver

Milton Kerker

Clarkson University, Potsdam, N.Y. 13676 USA

Abstract – The application of electromagnetic theory to account for the optical properties of colloidal dispersions has been extended to account for the remarkable enhancement of Raman signals from molecules adsorbed on colloidal silver. This review of work from the author's laboratory includes both theoretical and experimental studies.

It should come as no surprise that The Svedberg, who ranged so broadly through the field of colloid chemistry, should have also contributed to the optics of colloidal systems. In his PhD thesis (ref.1) Svedberg improved Bredig's (ref.2) method for preparing metal sols and with the aid of a newly constructed ultramicroscope he followed the Brownian motion and stability of these sols. Such sols, particularly those comprised of silver or gold particles, exhibit brilliant and varied colours.

Faraday (ref.3) had shown that preparations formed by reduction of gold salts were composed of elemental gold and Zsigmondy (ref.4) was able to observe, with the aid of the ultramicroscope, that this gold was in the form of colloidal particles. He noted that the colour depended upon particle size and shape and that these optical effects were similar to those observed for ruby glass. In 1908 Mie (ref.5) demonstrated that these varied colours could be accounted for by Maxwell's electromagnetic theory. We will see that there is a direct connection between the physical effect that gives rise to these colours and the subject of this lecture on surface enhanced Raman scattering (SERS).

Shortly after his PhD Svedberg published a study relating the absorption and colours of gold sols to the particle size (ref.6, 7) and this initial work was followed by a more comprehensive treatment by his student Pihlblad (ref.8) who was a pioneer in relating Mie's theoretical analysis to the spectrophotometry of these sols. Indeed, I recall reading Pihlblad's thesis in the course of reviewing the literature in preparation for my own doctoral research and marvelling at how pertinent and perceptive was his treatment.

THE COLOURS OF COLLOIDAL SILVER

Perhaps the most convenient measures of the optics of colloidal spheres are the efficiencies for extinction, scattering and absorption which are given by (ref.9)

$$Q_{ext} = (2/\alpha^2) \sum_{n=1}^{\infty} (2n+1)\{Re(a_n + b_n)\} \tag{1}$$

$$Q_{sca} = (2/\alpha^2) \sum_{n=1}^{\infty} (2n+1)\{|a_n|^2 + |b_n|^2\} \tag{2}$$

$$Q_{abs} = Q_{ext} - Q_{sca} \tag{3}$$

Q_{ext} represents the fraction of radiation incident on a particle which is scattered plus that which is absorbed within the particle. Q_{sca} and Q_{abs} are the corresponding quantities due to scattering alone and absorption alone. The size parameter is $\alpha = 2\pi a/\lambda$, where a is the radius and λ is the wavelength in the medium. The expansion coefficients a_n and b_n depend upon α and upon the complex refractive index relative to the medium $m = n(1-Ki)$ where K is the absorption index.

59

For very small particles ($\alpha \ll 1$; provided $\underline{m}$ is not too large, these expressions simplify to

$$Q_{sca} = (8\alpha^4)/3 \, |(\underline{m}^2-1)/\underline{m}^2+2)|^2 \tag{4}$$

$$Q_{abs} = -4\alpha \; \mathrm{Im}\{(\underline{m}^2-1)/\underline{m}^2+2)\} \tag{5}$$

It will be instructive for the purposes of this exposition to limit our treatment to this so-called Rayleigh limit, remembering that the complete analysis is accessible through eqn (1) to (3).

The colours of colloidal sols are determined by the wavelength dependence of Q_{sca} and Q_{abs}, i.e. the scattering and absorption spectra and the particular brilliance of gold and silver sols stems from the remarkable dispersion of the refractive index of these metals as illustrated in Table 1. Note for example that at 382 nm the refractive index of silver in water is approximately $\sqrt{2i}$ so that the denominator in eqn (4) and (5), ($\underline{m}^2+2$), becomes very small and the scattering and absorption become very large. (The expression never really blows up since there are higher order terms in the full theory which prevent this.)

The nearly purely imaginary value for the refractive index indicates that a silver particle can be envisaged as a cavity in which the balance of forces between the free electrons and the fixed cations is such that the electrons oscillate in phase with the exciting electromagnetic field. Any such cavity will have a particular frequency at which it will resonate and for spheres that frequency corresponds to the condition that $\underline{m} = \sqrt{2i}$. The resonance is characterized by extremely strong local fields which result in strong absorption and scattering. The strong absorption band at 382 nm for small silver spheres in water, as shown by the full curve in Fig. 1, accounts for the bright yellow colour of such a sol. The other curves in the figure are for equivolume prolate spheroids whose absorption bands move to successively longer wavelengths. The colour changes accordingly. The colours of such sols change in still different ways as the particles, whether spherical or of other shapes, increase in size.

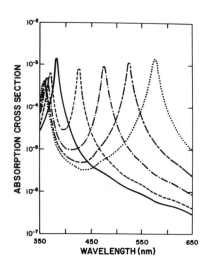

Table 1.
Complex refractive index of Au and Ag (ref. 33)

λ(nm)	Au	Ag	Ag (in H_2O)
354	1.50-1.87i	0.10-1.42i	0.08-1.07i
382	1.46-1.93i	0.05-1.86i	0.04-1.40i
413	1.47-1.95i	0.05-2.07i	0.04-1.56i
450	1.38-1.91i	0.04-2.66i	0.03-2.00i
496	1.04-1.83i	0.05-3.09i	0.04-2.32i
549	0.43-2.46i	0.06-3.59i	0.05-2.70i
617	0.21-3.27i	0.06-4.15i	0.05-3.12i
659	0.14-3.70i	0.05-4.48i	0.04-3.37i

In H_2O divide values by 1.33.

Fig. 1. Extinction cross section for a small silver sphere in water (solid curve) and for prolate spheroids with axial ratios 1.5, 2.0, 2.5 and 3.0 (in order from left to right).

SURFACE ENHANCED RAMAN SCATTERING (SERS)

We now turn to what may at first appear to be quite a different subject, surface enhanced Raman scattering. Raman scattering has not been a part of the vast array of surface sensitive spectroscopic techniques developed in recent years, even though it offers great advantages for determining the identity of adsorbed chemical species. The problem has been that the intensity of normal Raman scattering from a monolayer adsorbed on a macroscopic surface is much too low for practical detection.

Fleischmann, Hendra and McQuillan (ref.10) in England, in a series of papers starting in 1974 sought to overcome this limitation by working with pyridine adsorbed at a silver electrode surface which had been roughened by successive oxidation-reduction cycles. The main idea was that silver has a very high adsorptivity, that pyridine has a high Raman cross section, and that the increased surface area of the roughened surface would permit adsorption of a still larger number of molecules. Indeed, they obtained remarkably strong Raman signals but mistakenly attributed these to the increased surface concentration of pyridine at the roughened surface.

It was not until 1977 that Jeanmaire and Van Duyne (ref.11) in the United States and Albrecht and Creighton (ref.12) in England demonstrated that the signals were enhanced by a factor roughly estimated to be from 10^5 to 10^6 over what would be expected from a monolayer and that such a remarkable result could in no way be explained by an increase in surface area. It was a new phenomenon.

This finding unleashed a torrent of investigations that has hardly abated (ref.13, 14). Suitable substrates for adsorbed molecular species, in addition to the roughened electrodes on which this remarkable phenomenon was first observed, include colloidal metal particles, vacuum-deposited metal island films, matrix-isolated metal clusters, roughened surfaces of single crystals under ultrahigh vacuum, tunnel junction structures, smooth metal surfaces in the attenuated total reflection arrangement, metal-capped polymer posts, and holographic gratings. The initial observation of SERS for adsorbates on gold, silver, and copper have been extended to include aluminum, cadmium, lithium, nickel, palladium, platinum, and sodium.

Speculations to account for this surface-enhanced Raman scattering (SERS) immediately fell into two major categories. A purely physical mechanism was proposed in which the molecules were presumed to respond to gigantic electromagnetic fields generated locally by collective oscillations of the free electrons in small metal structures. In addition, so-called "chamical" mechanisms envisaged charge transfer between metal and adsorbate or else formation of a molecule-metal atom complex with consequent molecular resonances. Such specific molecular interactions may certainly play a role since different molecules on the same surface or different Raman bands of the same molecule may exhibit different SERS effects. Indeed, contributions to SERS from each of these two kinds of mechanisms, electromagnetic and chemical, are not mutually exclusive, yet the predominant current view is that the major contribution is electromagnetic and is due to the local field enhancement associated with resonant excitation of electron oscillations, otherwise termed surface plasmons.

Although the vast majority of experimental studies have utilized other substrates, colloidal particles are uniquely advantageous for theoretical analysis. Colloid optics is a venerable subject and we have expanded upon classical light scattering theory in order to articulate a complete electromagnetic field theory of SERS. Not only does this predict the magnitude of the enhancement in agreement with measurements, but also it depicts the remarkable wavelength dependence of SERS upon the optical properties and the morphology of the colloidal particles. The model can be considered as prototypic of the other substrates for none of which has it been possible to derive a definitive theory. Indeed, a substrate such as a "roughened" surface cannot even be described deterministically even though it can be very crudely envisaged as comparable to a smooth surface covered with "colloidal bumps".

ELASTIC AND INELASTIC LIGHT SCATTERING

We first consider elastic light scattering, for which there is no frequency shift. When a particle, composed of a homogeneous medium characterized by a refractive index $\underline{m}_o$ (Fig. 2), is irradiated by an electromagnetic wave at frequency ω_o, there is a field outside the particle constructed by superposition of the incident field $\underline{E}_o (\omega_o)$ plus a scattered field $\underline{E}_s (\omega_o, \underline{r}_>)$ and an internal field $\underline{E}_i (\omega_o, \underline{r}_<)$. These fields can be completely described in terms of the particle morphology and refractive index by appropriate matching at the particle boundary (ref.9).

In inelastic scattering by molecules, of which fluorescence and the Raman effect are examples, the frequency of the scattered light differs from that of the incident light. As a first approximation, the molecule is treated as a polarizable electric dipole which is excited at one frequency and re-emits at some shifted frequency.

There are several important areas of experimental investigation where inelastically scattering molecules are embedded within small particles, particularly Raman microprobe analysis of colloids and flow fluorimetry of biological cells and chromosomes. A number of

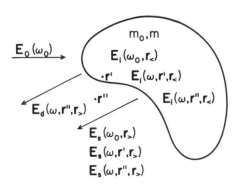

Fig. 2. Model for scattering by particle with refractive indexes $\underline{m}_o$, $\underline{m}$ at incident and Raman circular frequencies ω_o and ω. Raman dipoles at $\underline{r}'$ and $\underline{r}''$. Incident, internal, and scattered fields at ω_o are $\underline{E}_o (\omega_o)$, $\underline{E}_i (\omega_o)$, and $\underline{E}_s (\omega_o)$. Raman internal and scattered fields due to dipoles at $\underline{r}'$ and $\underline{r}''$ are $\underline{E}_i (\omega, \underline{r}')$, $\underline{E}_s (\omega, \underline{r}')$ and $\underline{E}_i (\omega, \underline{r}'')$, $\underline{E}_s (\omega, \underline{r}'')$ resp. $\underline{r}_<$ and $\underline{r}_>$ denote positions inside and outside the particle, respectively.

years ago we raised the following query regarding these experiments: How are inelastic signals affected by the embedding of active molecules within small particles? The result has been to show that such signals do indeed depend upon the particle morphology, the refractive index of the particle, and the spatial distribution of the active molecules within the particle.

To understand why this is so, we turn again to Fig. 2. An inelastic scattering molecule, located at some position $\underline{r}'$ within the particle, is excited by the radiation field at the incident frequency inside the particle (ref.15) and it will radiate at the shifted frequency, ω, for which the refractive index is $\underline{m}$. The Raman field $\underline{E}_i$ $(\omega, \underline{r}', \underline{r}_<)$ at the shifted frequency   inside the particle is composed of a dipolar field plus an induced field necessary to satisfy the boundary conditions. The Raman scattered field outside of the particle, $\underline{E}_s(\omega \underline{r}', \underline{r}_>)$, is obtained from the solution of this boundary value problem. For an array of molecules it is necessary to add the results appropriately (coherently or incoherently) to correspond to the particular distribution of Raman active material within the particle.

We have formulated the theory and carried out numerical calculations for spheres (ref. 16-19), concentric spheres (ref.20), cylinders (ref.21), and spheroids (ref.22, 23). Some of the main predictions of the theory have been verified by experiments with fluorescent polymer latexes (ref.24-29).

MODEL FOR SERS

The path has been straightforward, at least conceptually, from the theory of elastic scattering by small particles to the above model of inelastic scattering by molecules embedded within small particles. Extension to the case where the active molecules are outside of the particle, including positions at the outer surface, corresponds to what one finds in SERS.

The molecule (Fig. 2), once again presumed to behave as an electric dipole, is located at a position $\underline{r}''$ outside of the particle. The electric field outside the particle at frequency ω_o is comprised of the field of the incident plane wave $\underline{E}_o$ (ω_o) plus the scattered field $\underline{E}_s$ $(\omega_o, \underline{r}_>)$. This field excites the dipole located at $\underline{r}''$ to radiate at the Raman frequency. The Raman radiation field outside of the particle is comprised of the dipole field plus an induced field necessary to satisfy the boundary conditions. In this instance the induced field is the Raman scattered field for which the essential boundary value problem had been solved by us in another context (ref.30).

The enhancement is obtained by comparison of the Raman signal in the presence of the particle with that from a molecular dipole having the same polarizability as the surface-enhanced dipole but in the absence of the particle.

There is also an effect arising from the orientation of the various dipoles. In practice, our calculations have been performed by comparing Raman signals from a particle uniformly covered by dipoles whose axes are normal to the surface with a similar arrangement of dipoles in the solution. We assume that the absorption and emission dipoles are similarly oriented but uncorrelated in phase. In the case of spheroids, the particles are taken to be randomly oriented. There would be differences for other orientations of the dipoles at the surface, but we have not explored this in detail.

While these electromagnetic considerations provide the necessary framework for the model, they can be enriched by any specific chemical information available from either experiment or theoretical analysis. This may be done by incorporating that information into the Raman polarizability. The completely general solution for homogeneous spheres and the numerical results for that case will be discussed in the next section.

CALCULATION OF SERS FOR SPHERICAL PARTICLES

Our calculations for spherical particles give the dependence of the angular distribution and polarization of the SERS upon the size and dielectric properties of the particle as well as on the orientation of the molecular dipoles and their distance from the surface. The mathematical analysis and more detailed calculations are given elsewhere (ref.31, 32). The sampling of numerical results presented here is intended merely to illustrate some of the main physical features of the phenomenon.

Figure 3 depicts the enhancement as a function of excitation wavelength, i.e., the excitation profile, for a monolayer at the surface of 5, 50, and 500 nm-radius silver spheres immersed in water. The optical constants of silver (ref.33) are used because they exhibit the most striking effects. Indeed, that is why silver has been used in most of the experimental work. The Raman shift has been selected at 1010 cm^{-1} to correspond to this much studied pyridine line. The upper value of particle radius $\underline{a}$ = 500 nm has been dictated by the computational time requirements rather than by any limitations of the theoretical analysis.

The magnitude of the double-peaked maximum for the 5 nm particle agrees with the largest measured values (10^5-10^6). For $\underline{a}$ = 50 nm, the enhancement maximum of 10^4 is much broader and is shifted to longer wavelengths. For the still larger particle, $\underline{a}$ = 500 nm, the enhancement oscillates in the 10-10^2 range throughout the visible region.

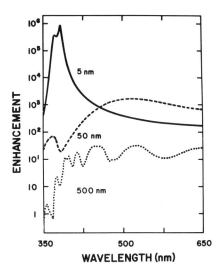

Fig. 3. SERS excitation profile for a
monolayer on 5, 50, and 500 nm-radius
Ag spheres in water for 1010 cm^{-1}
Raman band.

The peak value of 10^6 occurs only at about 382 nm and only for particles with radii
about 10 nm and smaller. The origin of this huge enhancement can be envisaged if we consider a very much simplified expression for the enhancement in the limiting case that the
particle be very much smaller than the wavelength viz.

$$\underline{G} = |(1 + 2\underline{g})(1 + 2\underline{g}_0)|^2 \qquad (6)$$

$$\underline{g} = \frac{\underline{m}^2 - 1}{\underline{m}^2 + 2} \qquad \underline{g}_0 = \frac{\underline{m}_0^2 - 1}{\underline{m}_0^2 + 2} \qquad (7)$$

The origin of the enhancement is now apparent. It occurs because of the very same resonance
that gives rise to the strong absorption and the colour effects for colloidal silver, i.e.,
the condition that $\underline{m} \rightarrow \sqrt{2i}$. However, in this case there may be two coupled resonances; one
at the incident frequency ω_0 and one at the Raman shifted frequency ω provided the Raman
shift is sufficiently small so that both frequencies will be sufficiently close to
resonance.

This connection between SERS on the one hand and the absorption and scattering spectra
on the other is shown in Fig. 4 where the SERS excitation profile, the absorption spectrum,
and the scattering spectrum of a 5 nm-radius silver sphere in water are compared. Note how
each of these peaks is at precisely the same excitation wavelength.

For larger spheres there will, in addition to the dipolar mode, be a hierarchy of
higher order modes all of which contribute to the various optical processes associated with

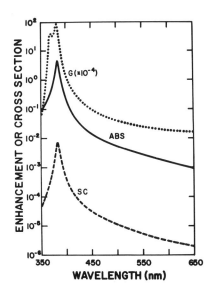

Fig. 4. Comparison for 5 nm-radius Ag sphere
in water of the SERS excitation profile
(...) for 1010 cm^{-1}, absorption cross
section (———), and scattering cross section
(____).

the spherical particles. Any one of these may be resonantly excited at a particular
frequency which depends upon particle size and optical constants. Actually, resonant
excitation of a particular mode may result in very strong fields localized at radial
positions inside the particle, thereby giving rise to strong absorption, whereas resonant
excitation of some other mode may give rise to strong fields localized just outside the
particle surface, giving rise to SERS. Accordingly, for larger spheres the absorption and
scattering spectra will be decoupled from the SERS excitation spectrum. This point should
be emphasized. The absorption and scattering spectra are necessarily coupled with the SERS
excitation profile only when the particles are sufficiently small to be treated in the
dipole limit.

SPHEROIDS

The most striking feature of SERS is that the gigantic enhancement occurs over a narrow
wavelength region and that for spheres much smaller than the wavelength this excitation
profile parallels a sharp band in the absorption and scattering spectra. These effects are
sensitive to particle morphology and we now extend the model to prolate spheroids. The
treatment is restricted to the limit that the particle be small compared to the wavelength.
An ellipsoidal particle radiates as if it were a polarizable dipole with polarizability
$(\underline{abc})\underline{g_i}$ where

$$\underline{g_i} = \frac{(\underline{m}^2-1)}{\underline{3}\left[1+(\underline{m}^2-1)\underline{P_i}\right]} \qquad\qquad \underline{i} = \underline{a},\underline{b},\underline{c} \qquad\qquad (8)$$

and the depolarization factor $\underline{P_i}$ depends upon the values of the three semiaxes of the
ellipsoid; $\underline{a}$, $\underline{b}$, and $\underline{c}$ (ref. 9). For a sphere, the triply degenerate values of $\underline{P_i} = 1/3$ so
that resonance, as we have already seen, occurs whenever $\underline{m}^2 = -2$. For an ellipsoid there
may be three distinct absorption bands corresponding to the three values of $\underline{P_i}$; for a
spheroid there may be two bands, one of which will be doubly degenerate.
     This dependence of the optical absorption of small particles on shape has been known
for a long time. We have already seen (Fig. 2) that as the shape of an equivolume silver
sphere is changed to increasingly elongated prolate spheroids, the triply degenerate absorption band at 382 nm for the sphere splits into two bands, which depend upon the two semiaxes
of the spheroids, one of these at a lower wavelength and a doubly degenerate one at a higher
wavelength. For a 3 to 1 axial ratio, the longer wavelength band has shifted to 580 nm. In
the parlance of the resonance model there is a dipolar surface plasmon resonating at each of
these wavelengths.
     It is not surprising that there is a similar effect upon the SERS excitation profile.
We have utilized the same physical model as for spheres but because of the added
mathematical complexities, have only obtained a result valid for particles which are small
relative to the wavelength (ref.34). Somewhat similar results were also obtained by
Gersten and Nitzan (ref.35) and by Adrian (ref.36). A typical calculation is shown in
Fig. 5, where the various curves represent the excitation profiles for a silver sphere in
aqueous medium and for equivolume spheroids of increasing elongation. The molecular dipoles
are assumed to form a monolayer and to be oriented normal to the surface, and the particles
are assumed to be randomly oriented. Not only does the strongly enhanced band shift but the

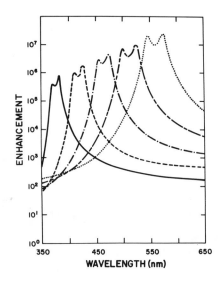

Fig. 5. SERS excitation profiles for
monolayer on a small silver sphere in
water (solid curve) and for prolate
spheroids with axial ratios 1.5,
2.0, 2.5 and 3.0 (in order from
left to right). Raman shift is 1010 cm^{-1}.

enhancement itself increases by an order of magnitude. The longer wavelength branch of the bimodal peak occurs at the same wavelength as the corresponding peak in the absorption spectrum. The separation of the two modes in the SERS band equals the Raman shift, in this instance 1010 cm^{-1}.

There are a number of other aspects of our theoretical analysis which will not be treated here. For example, whenever one of the two regions of a particle consisting of a spherical core and a concentric spherical shell is composed of silver, the silver region behaves as a resonant cavity whose resonance condition depends upon the shell thickness. Enhancement occurs whether the Raman molecules are adsorbed at the outer surface (ref.37, 38) or embedded within the dielectric region (ref.39, 40). Other studies include the effect of dipole-dipole interaction as a result of varying surface coverage (ref.41), an electro-optical effect due to surface charge (ref.42), the dependence of SERS upon the magnitude of the Raman shift (ref.43), solvent effects (ref.43), and analogous enhancement effects upon other optical processes such as coherent anti-Stokes Raman scattering (CARS) (ref.44). Instead we now turn to some experimental considerations.

EXPERIMENTS

The strong dependence of the enhanced Raman signal upon the incident wavelength is shown in Fig. 6 where at the right the 1400 cm^{-1} band of citrate adsorbed on the silver particles of a Carey Lea hydrosol for several excitation wavelengths is compared with the corresponding quantities for a 1.37 M sodium citrate solution (ref.45). A radiotracer technique (ref. 46) was used to determine that the concentration of adsorbed citrate was only 6×10^{-6} M so that in the visible region the peak enhancement was between 10^5 to 10^6 in agreement with theory.

However, just as the absorption spectrum of the sol depends upon the particular distribution of particle size and shape, so does the SERS excitation spectra. This is illustrated in Fig. 7 where the absorption (extinction) spectra and the corresponding SERS excitation are compared for four sol preparations with varying degrees of aggregation. The fraction of aggregates, which is quite small in sol 1 and increases through sol 4, is manifested by the development of a shoulder at 490 nm in the absorption curve which develops for sol 4 into a full-fledged second maximum at 550 nm. The colour of the sol changes from yellow to deep red. The SERS peaks occur at the corresponding wavelength. Indeed, qualitative agreement with these results is obtained when the doublets and triplets comprising the aggregates are modelled by equivolume spheroids (ref.47). We are presently attempting to extend the theoretical analysis to doublets and triplets to provide a more definitive comparison between theory and experiment.

Whenever the incident radiation is tuned into an electronic absorption band of a molecule, the Raman signal is strongly enhanced, an effect called resonance Raman scattering. We selected dabsylaspartate as a chromophoric molecule because its absorption band overlapped with that of the silver hydrosol offering the possibility of combined surface and resonant enhancement. Indeed, enhancements as large as 10^9 were observed (ref.

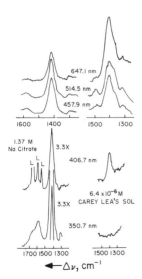

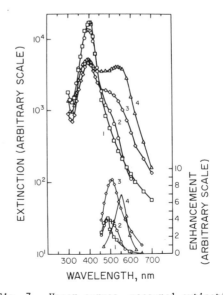

Fig. 6. Raman bands for excitation at λ$_9$ = 350.7, 406.7, 457.9, 514.5 and 647.1 nm. Left hand side 1.37 M sodium citrate. Right hand side, Carey-Lea silver hydrosol.

Fig. 7. Upper curves, measured extinction spectra for four silver hydrosols. Lower curves, corresponding measured SERS excitation profiles (arbitrary units). The sols are increasingly aggregated from 1 to 4.

48). The overlap could be varied by utilizing sols having different degrees of aggregation or by utilizing a dispersion of colloidal silver in ethanol (ref.49). In this way the respective contributions of resonance Raman mechanism and the electromagnetic theory mechanism articulated here could be elucidated.

REFERENCES

1. T. Svedberg, Nova Acta R. Soc. Sci. Uppsala (4) 2, No. 1, 1 (1907).
2. G. Bredig, Anorganische Fermente, Wilhelm Engelmann, Leipzig, 1901.
3. M. Faraday, Phil. Trans. 147, 145 (1857).
4. R.A. Zsigmondy, Zur Erkenntnis der Kolloide, Gustav Fischer, Jena, 1905.
5. G. Mie, Ann. Phys 25, 377 (1908).
6. T. Svedberg, Z. Phys. Chem. 65, 624 (1909).
7. T. Svedberg, Z. Chem. Ind. Kolloide 4, 168 (1909).
8. N. Pihlblad, Lichtabsorption und Teilchengrösse in Dispersen Systemen, Almqvist and Wiksell, Uppsala (1918).
9. M. Kerker, The Scattering of Light and other Electromagnetic Radiation, Academic Press, New York (1969).
10. M.J. Fleischmann, P.J. Hendra and A.J. McQuillan, Chem. Phys. Lett. 26, 163 (1974).
11. D.J. Jeanmaire and R.P. Van Duyne, J. Electroanal. Chem. 84, 1 (1977).
12. M.G. Albrecht and J.A. Creighton, J. Am. Chem. Soc. 99, 5215 (1977).
13. R.K. Chang and T.E. Furtak, Surface Enhanced Raman Scattering, Plenum Press, New York (1982).
14. A. Otto, in Light Scattering in Solids, Vol. IV, eds. M. Carbona and G. Guntherodt, Springer, New York (1983).
15. P.W. Dusel, M. Kerker and D.D. Cooke, J. Opt. Soc. Am. 69, 55 (1979).
16. H. Chew, P.J. McNulty and M. Kerker, Phys. Rev. A 13, 396 (1976).
17. M. Kerker, P.J. McNulty, M. Sculley, H. Chew and D.D. Cooke, J. Opt. Soc. Am. 68, 1676 (1978).
18. H. Chew, M. Sculley, M. Kerker, P.J. McNulty and D.D. Cooke, J. Opt. Soc. Am. 68, 1686 (1978).
19. M. Kerker and S.D. Druger, Appl. Opt. 18, 1172 (1979).
20. H. Chew, M. Kerker and P.J. McNulty, J. Opt. Soc. Am. 66, 440 (1976).
21. H. Chew, D.D. Cooke and M. Kerker, Appl. Opt. 19, 44 (1980).
22. D.-S. Wang, M. Kerker and H. Chew, Appl. Opt. 19, 2315 (1980).
23. M. Kerker, D.-S. Wang and H. Chew, Cytometry 1, 161 (1980).
24. J.P. Kratohvil, M.P. Lee and M. Kerker, Appl. Opt. 17, 1978 (1978).
25. P.J. McNulty, S.D. Druger, M. Kerker and H. Chew, Appl. Opt. 18, 1484 (1979).
26. M. Kerker, M. van Dilla, A. Brunsting, J.P. Kratohvil, P. Hsu and D.-S. Wang, Cytometry 3, 71 (1982).
27. E.-H. Lee, R.E. Benner, J.B. Fenn and R.K. Chang, Appl. Opt. 17, 1980 (1978).
28. R.E. Benner, R. Bornhaus, M.B. Long and R.K. Chang, in Microbeam Analysis, ed. D.E. Newberry, San Francisco Press, San Francisco (1979).
29. R.E. Benner, J.F. Owen and R.K. Chang, J. Phys. Chem. 84, 1602 (1980).
30. H. Chew, M. Kerker and D.D. Cooke, Phys. Rev. A 16, 320 (1977).
31. D.-S. Wang, H. Chew and M. Kerker, Appl. Opt. 19, 2256 (1980).
32. M. Kerker, D.-S. Wang and H. Chew, Appl. Opt. 19, 4159 (1980); this is a rerun of the version published on p. 3373 in which the many typographical errors were not amended.
33. P.B. Johnson and R.W. Christy, Phys. Rev. B 6, 4370 (1972).
34. D.-S. Wang and M. Kerker, Phys. Rev. B 24, 1777 (1981).
35. J. Gersten and A. Nitzan, J. Chem. Phys. 73, 3023 (1980).
36. F.J. Adrian, Chem. Phys. Lett. 78, 45 (1981).
37. A.L. Aden and M. Kerker, J. Appl. Phys. 22, 1242 (1951).
38. M. Kerker and C.G. Blatchford, Phys. Rev. B 26, 4052 (1982).
39. D.-S. Wang and M. Kerker, Phys. Rev. B 25, 2433 (1982).
40. M. Kerker and D.-S. Wang, Chem. Phys. Lett. 104, 516 (1984).
41. H. Chew, D.-S. Wang and M. Kerker, Phys. Rev. B 28, 4169 (1983).
42. C.G. Blatchford, O. Siiman and M. Kerker, J. Phys. Chem. 87, 2503 (1983).
43. C.G. Blatchford, M. Kerker and D.-S. Wang, Chem. Phys. Lett. 100, 230 (1983).
44. H. Chew, D.-S. Wang and M. Kerker, J. Opt. Soc. Am. B 1, 56 (1984).
45. M. Kerker, O. Siiman, L.A. Bumm and D.-S. Wang, Appl. Opt. 19, 3253 (1980).
46. O. Siiman, L.A. Bumm, R. Callaghan, C.G. Blatchford and M. Kerker, J. Phys. Chem. 87, 1014 (1983).
47. M. Kerker, O. Siiman and D.-S. Wang, J. Phys. Chem. 88, 3168 (1984).
48. O. Siiman, A. Lepp and M. Kerker, J. Phys. Chem. 87, 5319 (1983).
49. O. Siiman, A. Lepp and M. Kerker, Chem. Phys. Lett. 100, 163 (1983).

11. Colloidal Behaviour of Surfactant Systems: Micelles, Microemulsions and Liquid Crystals

Björn Lindman and Håkan Wennerström

Physical Chemistry 1, Chemical Centre, Lund University, S-221 00 Lund, Sweden

Abstract – Some current research topics in the field of surfactant systems are reviewed. In particular the following recent results are discussed:

- Lamellar liquid crystalline phases of surfactants with divalent counterions have a factor of ten lower capacity to take up water than those of surfactants with monovalent counterions. This important difference between for example Ca^{2+} and Na^+ can be referred to electrostatic effects including counterion/counterion correlations.
- Hydrocarbon chain order in micelles is close to that of liquid crystals according to an analysis of multified NMR relaxation using a two-step motional model. Rapid internal motions (ca. 10^{-11} seconds) are similar to those of n-alkanes.
- The size of oligo(ethyleneoxide) surfactant micelles is very sensitive to temperature and concentration. Micelle growth can be rationalized in terms of intra- and inter-micellar interactions determined by a temperature-dependent headgroup hydration.
- Self-diffusion studies of various microemulsion systems reveal a high microstructure variability with oil-in-water droplet, water-in-oil droplet and bicontinuous structures depending on cosurfactant, surfactant, added salt and temperature.

INTRODUCTION

Amphiphilic molecules like surfactant molecules, which are built up of one large hydrophobic part and, rather distinctly separated from this, one strongly hydrophilic group, show in aqueous systems a strongly cooperative self-association into aggregates of colloidal dimensions (ref.1-6). This self-association can, depending on conditions (such as surfactant chemical structure, concentration, temperature etc.), lead to systems which are macroscopically extremely different (e.g. rheological and optical properties); often quite subtle changes can bring about major changes in macroscopic behaviour. In contrast to this, one observes on the molecular level only minute changes in properties and one important task of current research is to elucidate the relation between molecular effects and macroscopic properties, which are so important for a number of biological systems and for a manifold of technical applications.

Common to the different aggregates in surfactant – water systems is a quite sharp separation into hydrophilic and hydrophobic domains. For ionic surfactant systems, the high charge densities of the aggregates have important influences on micelle size and shape and on phase diagrams. The micelle/micelle interaction is, furthermore, repulsive under a wide range of conditions (temperature, concentration, nonelectrolyte additives). A consequence of this is that for ionic surfactant micelles, the micelles are well-defined entities with fusion as a rather improbable event.

For nonionic surfactant systems, the pattern is often much more complex and depending on surfactant headgroup, temperature and concentration both repulsive and attractive situations are possible. An important group of nonionic surfactants are those where the polar group is an oligo(ethyleneoxide) chain. Here the behaviour has been found to be sensitive to quite small changes in temperature, concentration and additives.

In three-component systems of surfactant, water and an essentially water-insoluble substance like a hydrocarbon or a long-chain alcohol, or in four-component systems of water and hydrocarbon as well as two surface-active substances (a normal surfactant and a cosurfactant), there are often quite large concentration domains of isotropic thermodynamically stable solutions. These are generally termed microemulsions and are currently attracting a lot of interest. While from an applied point of view, systems with low surfactant concentrations and high concentrations of both hydrocarbon and water are of particular relevance, most investigations are because of methodological limitations carried out on other regions of the phase diagram (low amounts of water or hydrocarbon, high surfactant/cosurfactant concentration).

In this article we will briefly summarize some current topics of research in our laboratory dealing with different aspects of surfactant self-association.

PHASE BEHAVIOUR OF CALCIUM SURFACTANTS

Not least through the pioneering work of Ekwall and coworkers, the phase behaviour of surfactants with monovalent counterions (mainly sodium but to some extent also chloride, bromide, potassium and lithium) has been very extensively mapped (ref.7, 8). Three-component systems of these surfactants generally give rise to two isotropic solution phases and two or more liquid crystalline phases. In particular, a common feature of many of these phase diagrams is an extensive region of lamellar liquid crystalline phase, built up of alternating layers of water and surfactant bilayers.

Despite their biological and practical relevance, very little work has been done with ionic surfactants with a divalent counterion such as calcium or magnesium. Work (ref.9-11) was started a few years ago by Ali Khan and his colleagues to map systematically the phase behaviour and association properties of divalent counterion surfactants. This project has now started to provide evidence for a common behaviour of these surfactants which is clearly different from the monovalent case.

As illustrated in Fig. 1 the most striking difference lies in the stability region of the lamellar phase. With monovalent counterions the lamellar mesophase generally has a strong tendency to take up water and swell while very little water can be incorporated when calcium and magnesium are counterions. The difference is typically as large as a factor of ten in the maximal number of water molecules per surfactant molecule.

Since the swelling behaviour of ionic surfactant systems is dominated by electrostatic interbilayer repulsions (ref.12) determined by the counterion distribution, the difference

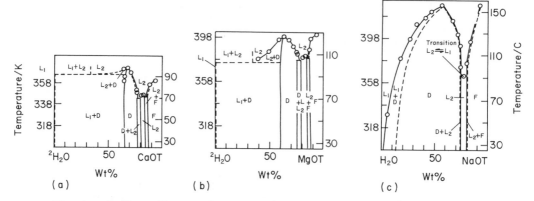

Fig. 1. A. Phase diagrams for some sodium, magnesium and calcium surfactant systems. (From refs. 9-11.)

(a) Two-component metal ion di-2-ethylhexylsulphosuccinate/water (D_2O) systems. (b) shows the order parameter and (c) the fast motion correlation time along the alkyl chain. The slow motion correlation time was found to be 2 nanoseconds eqn (from ref.19).

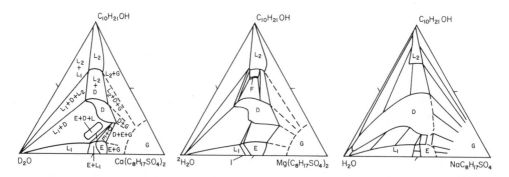

Fig. 1. B. Three-component metal ion octylsulphate/decanol/water (D_2O) systems at 25°C. The region of lamellar phase of special concern in this article is labelled D. L_1 and L_2 are isotropic solution phases and E, F and I_2 other liquid crystalline phases. All concentrations in weight per cent. For further details see original papers (refs.7-15).

in lamellar phase stability is explained by a much larger tendency of divalent than of monovalent counterions to reside close to the charged surfaces. This difference in counterions "binding" is for the micellar solutions borne out in direct experimental observations (conductivity, self-diffusion) (ref.13). Electrostatic Poisson/Boltzmann theoretical calculations qualitatively predict the observed behaviour but, neglecting counterion/counterion correlations, tends to underestimate the effect which is more precisely obtained in computer simulations of the Monte Carlo type (ref.14, 15).

HYDROCARBON CHAIN ORDER AND DYNAMICS IN MICELLES

It was already realized in early studies (ref.16) of surfactant micelles that the micelle interior has a "liquid-like" character and thus can be regarded roughly as a microscopic oil droplet. However, due to experimental limitations a quantification of hydrocarbon chain order and dynamics has been slow to emerge. For anisotropic liquid crystals, the situation is better since hydrocarbon chain order can be directly measured by NMR methods. Mainly ^2H quadrupole splittings have provided quantitative data on the order in the chains.

For isotropic systems the realization that different motions occur on distinctly different time-scales provided a useful way of monitoring chain order and dynamics via multi-frequency NMR relaxation experiments (ref.17). In the "two-step" model of relaxation, part of the interaction is considered to be modulated by rapid local motions while the remainder is averaged by motions over the extension of the surfactant aggregate (e.g. micelle rotation). Very extensive tests of the model by Olle Söderman and his colleagues have verified that it is indeed possible to distinguish between rapid internal motions (typically 10^{-11} seconds) and slow overall motions (10^{-9} seconds and upwards). From multi-field ^{13}C or ^2H NMR relaxation experiments it is possible to extract (for each methylene or methyl group) three quantities, i.e. an order parameter (S), describing the degree of orientation of a molecular vector with respect to the micelle surface and the characteristic correlation times of the slow (τ_c^s) and fast (τ_c^f) motions (ref.18).

From the results for several micellar systems (as exemplified in Fig. 2) the following general conclusions emerge:

a) The motions in the hydrocarbon chains are very rapid and in fact the fast motion correlation times are similar to those of n-alkanes with the same number of carbon atoms.

b) The order parameters are similar in magnitude to those of lamellar liquid crystals but the variations along the alkyl chains are somewhat different.

c) The slow motion correlation times are influenced both by micelle rotation and surfactant molecule lateral diffusion along the micelle surface. Correlation times are in qualitative but not quantitative agreement with results from simple geometric considerations for spherical micelles.

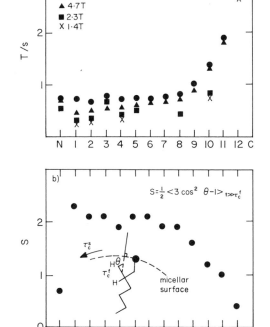

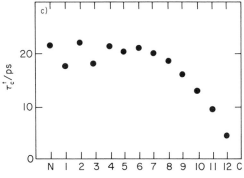

Fig. 2. ^{13}C NMR spin relaxation data (a) at different magnetic fields for dodecyltrimethylammonium micelles at 27°C as well as results from an analysis in terms of a two-step model of relaxation (from ref.18).

OLIGO(ETHYLENEOXIDE) SURFACTANT MICELLES

The great variability in the phase behaviour of aqueous oligo(ethyleneoxide) alkyl ether (often abbreviated C_xE_y, $\underline{x}$ being the number of carbon atoms in the alkyl chain and $\underline{y}$ the number of ethyleneoxide groups) systems is well documented in previous work (see refs. 20 and 21 and papers cited therein). These surfactants show liquid crystal formation which is strongly dependent on surfactant chemical structure, and as a rather common feature a phase separation (so-called clouding phenomenon) into two isotropic solutions for quite dilute systems on an increase in temperature. In a recent thesis from our group by Per-Gunnar Nilsson the micellar solutions were studied in some detail as regards micelle size, shape, hydration and intermicellar interactions for a few cases (mainly $C_{12}E_5$ and $C_{12}E_8$) (ref.21). Experimental parameters used were essentially water and surfactant self-diffusion and ^1H NMR relaxation.

The cmc's (critical micelle concentration) of these systems are very low so measured surfactant self-diffusions coefficients correspond in a broad concentration region to the micelle self-diffusion coefficient. The behaviour of the $C_{12}E_5$ and $C_{12}E_8$ systems is found to be strikingly different. The low concentration data (illustrated in Fig. 3) give micellar radii from the Stokes/Einstein relation which increase strongly with increasing temperature for $C_{12}E_5$ but which remain approximately constant at the minimal spherical size for $C_{12}E_8$. An equally striking difference is observed as regards the change of micelle size with increasing concentration, $C_{12}E_5$ micelles again growing very strongly while $C_{12}E_8$ micelles remain small up to very high concentrations (ref.23). These observations of growth of $C_{12}E_5$ micelles as well as the absence of growth $C_{12}E_8$ are also clearly documented in the ^1H relaxation work. There are indications that growth of $C_{12}E_5$ micelles results in rod micelles rather than discs (ref.21. 22).

Water self-diffusion in colloidal systems provides information (using a simple two-site model) on the number of water molecules diffusing with the colloidal particles and, therefore, on hydration. Studies of these surfactant systems demonstrate that hydration decreases with increasing concentration or temperature and increases with increasing number of ethyleneoxide groups (ref.24). A very striking observation is that hydration per ethyleneoxide group is approximately the same for quite different systems (rod or spherical micelles, polyethylene-glycol) as illustrated in Fig. 4.

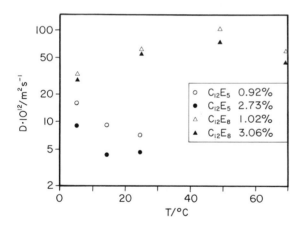

Fig. 3. $C_{12}E_5$ and $C_{12}E_8$ self-diffusion coefficients in dilute aqueous solutions as a function of temperature (from ref.22).

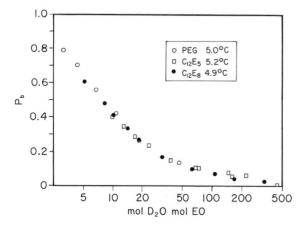

Fig. 4. Fraction of bound water, $\underline{P}_b$, $\underline{vs.}$ the molar ratio water (D_2O)/ethyleneoxide groups for polyethyleneglycol, $C_{12}E_5$ and $C_{12}E_8$ at $5°C$ (from ref.24).

The strong hydration of C_xE_y micelles affects both intra- and inter-micellar interactions. Strong hydration favours small aggregates with a high curvature from simple geometrical packing considerations. Therefore, the minimal spherical micelles are favoured by low temperature, low concentration and long EO-chains while growth would be expected for small polar heads especially at higher concentrations and/or temperature. A most striking observation is that such a relatively small change as between $C_{12}E_5$ and $C_{12}E_8$ so dramatically affects the behaviour. An illustration of this difference is given by the surfactant self-diffusion coefficients displayed in Fig. 5. As can be seen, $C_{12}E_5$ diffuses more than a factor of 10 faster than $C_{12}E_8$ at low concentrations, while the reverse is true at high concentrations. It is possible to analyse such data in terms of micellar growth and intermicellar interactions. $C_{12}E_8$ micelles are more strongly hydrated and stay essentially spherical throughout the entire concentration range. The hydration leads to a strong inter-micellar repulsion and thus to slow diffusion. For $C_{12}E_8$, this "hydration force" is weaker and intermicellar interactions between the big micelles become appreciably attractive at higher concentrations. This leads to less retardation of micelle diffusion and more significantly leads to a facile approach between micelles. This permits an intermicellar exchange of surfactant molecules between micelles, which for big micelles can have important effects on the diffusion coefficients even if the micelle/micelle encounters are relatively improbable.

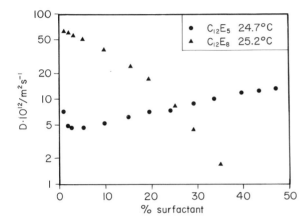

Fig. 5. Self-diffusion coefficients of $C_{12}E_5$ and $C_{12}E_8$ at 25°C vs. surfactant concentration (from ref.22).

MICROSTRUCTURE OF MICROEMULSIONS

The classical picture of microemulsion structure emerged from an assumption that it should be closely related to that of real (macro)emulsions which are two-phase systems where one phase is dispersed as drops in the other. One therefore considered oil-in-water and water-in-oil microemulsions with "droplet" structure and drops of the order of 100 Å or below in size. We had found self-diffusion studies useful for isotropic surfactant systems and demonstrated that confinement of a constituent into microdomains leads to a very substantial reduction in the translational mobility of the constituent over macroscopic distances. Together with B. Brun and coworkers in Montpellier we started self-diffusion work on four- and five-component microemulsion systems using the capillary tube method with radioactive labelling (ref.25). As these multi-component self-diffusion studies are very laborious and require isotope-labelled compounds it was somewhat of a revolution for this field when Stilbs in Uppsala developed his Fourier transform NMR pulsed-gradient spin-echo technique for the simultaneous determination of several components' self-diffusion coefficients (ref. 26). Thanks to this development, experimental data for a wide range of systems are now available.

A survey of a range of microemulsion systems demonstrates a very broad structural variability (ref.19, 27) as implied by a variation in the water-to-hydrocarbon diffusion, $\underline{D}_w/\underline{D}_{hc}$, by a factor of 10^5. Typical oil-in-water systems ($\underline{D}_w \gg \underline{D}_{hc}$) are exemplified by micellar solutions with solubilizates while a distinct water-in-oil structure ($\underline{D}_w \ll \underline{D}_{hc}$) is observed in systems of double-chain surfactant, hydrocarbon and water. "Classical" four-component microemulsions with a cosurfactant are structurally very dependent on cosurfactant and composition (ref.25, 28). Fig.6 demonstrates that for microemulsions with relatively high concentrations of all components the system may be distinctly of a W/O character with decanol as cosurfactant but bicontinuous with pentanol or butanol. These latter cases are typically bicontinuous over very wide concentration ranges but adopt more of a droplet-type character on approach of the high water or hydrocarbon concentration limits. The structural changes are very gradual.

For ionic surfactant systems, salinity is known strongly to affect the phase behaviour and Fig. 7 illustrates that salinity also has a very profound influence on microemulsion structure (ref.29, 30). While this system has an O/W character at low salinities it changes via a bicontinuous structure to W/O structure at high salinities.

In conclusion, multi-component self-diffusion appears to provide a facile insight into the structural variability of complex surfactant systems like microemulsions.

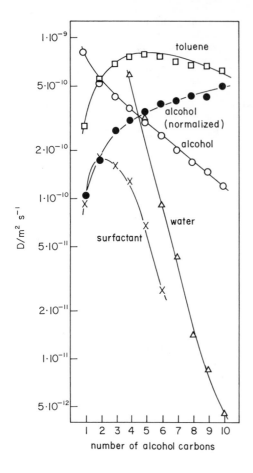

Fig. 6. Self-diffusion data, obtained with the FT NMR method on a series of alcohol/sodium dodecylsulphate/toluene/water system at a constant (35.0:17.5:12.5:35.0) weight fraction composition (from ref.19).

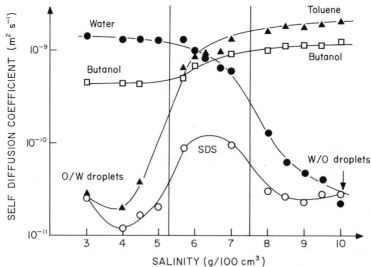

Fig. 7. Effect of salinity on constituent self-diffusion coefficients for microemulsions composed of (weight per cent): 46.25 % toluene, 46.8 % aqueous NaCl solution, 1.93 % sodium dodecylsulphate and 3.93 % butanol (from ref.30).

REFERENCES

1. K. Shinoda, T. Nakagawa, B.I. Tamamushi and T. Isemura, Colloidal Surfactants, Academic Press, New York (1963).
2. C. Tanford, The Hydrophobic Effect, John Wiley, New York (1980).
3. D. Attwood and A.T. Florence, Surfactant Systems, Chapman and Hall, London (1983).
4. Th.F. Tadros (ed.), Surfactants, Academic Press, London (1984).
5. H. Wennerström and B. Lindman, Phys. Rep. **52**, 1 (1979).
6. B. Lindman and H. Wennerström, Top. Curr. Chem. **87**, 1 (1980).
7. P. Ekwall, Adv. Liquid Cryst. **1**, 1 (1975).
8. P. Ekwall, L. Mandell and K. Fontell, Mol. Cryst. Liquid Cryst. 8, 157 (1969).
9. A. Khan, K. Fontell, G. Lindblom and B. Lindman, J. Phys. Chem. **86**, 4266 (1982).
10. A. Khan, K. Fontell and B. Lindman, J. Colloid Interface Sci. **101**, 193 (1984).
11. A. Khan, K. Fontell and B. Lindman, Colloids and Surfaces, in press.
12. B. Jönsson and H. Wennerström, J. Colloid Interface Sci. **80**, 482 (1981).
13. B. Lindström, O. Söderman, A. Khan and B. Lindman, 8th Scand. Symposium on Surface Chemistry, Lund, 1984.
14. H. Wennerström, B. Jönsson and P. Linse, J. Chem. Phys. **76**, 4665 (1982).
15. L. Guldbrand, B. Jönsson, H. Wennerström and P. Linse, J. Chem. Phys. **80** 2221 (1984).
16. G.S. Hartley, Aqueous Solutions of Paraffin-Chain Salts, Hermann, Paris, 1936.
17. H. Wennerström, B. Lindman, O. Söderman, T. Drakenberg and J. Rosenholm, J. Am. Chem. Soc. **101**, 6860 (1979).
18. H. Walderhaug, O. Söderman and P. Stilbs, J. Phys. Chem. **88**, 1655 (1984).
19. P. Stilbs and B. Lindman, Colloid & Polymer Sci., in press.
20. D.J. Mitchell, G.J.T. Tiddy, L. Waring, T. Bostock and M.P. McDonald, J. Chem. Soc. Faraday 1, **79**, 975 (1983).
21. P.G. Nilsson, Doctoral Thesis, Lund, 1984.
22. P.G. Nilsson, H. Wennerström and B. Lindman, Proc. Conf, on Hydration Forces and Molecular Aspects of Solvation, Örenäs, Chem. Scr. (1984), in press.
23. P.G. Nilsson, H. Wennerström and B. Lindman, J. Phys. Chem. **87**, 1377 (1983).
24. P.G. Nilsson and B. Lindman, J. Phys. Chem. **87**, 4756 (1983).
25. B. Lindman, N. Kamenka, T.M. Kathopoulis, B. Brun and P.G. Nilsson, J. Phys. Chem. **84**, 2485 (1980).
26. P. Stilbs and M.E. Moseley, Chem. Scr. **15**, 176 (1980).
27. B. Lindman and P. Stilbs, Proc. World Surfactant Congress, München, p. 159 (1984).
28. B. Lindman, T. Ahlnäs, O. Söderman, H. Walderhaug, K. Rapacki and P. Stilbs, Disc. Faraday Soc. **76**, 317 (1983).
29. D. Chatenay, P. Guèring, W. Urbach, A.M. Cazabat, D. Langevin, J. Meunier, L. Léger and B. Lindman, 5th Int. Sym. on Surfactants in Solution, Bordeaux (1984).
30. P. Guèring and B. Lindman, in preparation.

12. Separation of Cells and Other Particles by Centrifugation in Colloidal Silica Solutions

Håkan Pertoft and Torvard C. Laurent

Institute of Medical and Physiological Chemistry, Biomedical Centre, Uppsala University, Box 575, S-751 23 Uppsala, Sweden

Abstract - Stable density gradients can be formed from solutions of colloidal silica particles coated with polyvinylpyrrolidone. These gradients are well suited to prevent convection during rate sedimentation or isopycnic banding of tissues, cells and cell organelles. The material is non-toxic and the osmotic pressure, pH, ionic strength and ionic composition can be adjusted to physiological conditions.

This short review summarizes a project which is based on four of the main interests of The Svedberg, i.e. inorganic colloids, ultracentrifugation, application of physical chemistry to biology and medicine and utilization of academic discoveries in Swedish Industry.

As is often the case, our project arose out of a case of serendipity. In our studies on separation of virus particles by centrifugation in polysaccharide solutions (ref.1) we searched for a material, which could form density gradients in cylindrical centrifuge tubes. Brakke (ref.2) and Pickles (ref.3) had already in the early 1950s proposed the use of sucrose gradients to restrict convective movements in sedimentation experiments and Meselson, Stahl and Vinograd (ref.4) developed an elegant technique to band DNA isopycnically in gradients of CsCl. However, de Duve et al. eqn (ref.5) showed that biological particles, such as cells and cell organelles, change their hydration when centrifuged in e.g. sucrose or CsCl and thus also their size and density. The ideal gradient medium for biological particles should have a high density, low viscosity, moderate osmotic pressure and be miscible with water. By chance we had a bottle of colloidal silica on the shelf - used for calibration of a light-scattering photometer - and we decided to try this colloid to make density gradients. It became apparent from the beginning that gradients of colloidal silica had interesting properties. After a decade of development a modified colloidal silica was introduced commercially for separation of biological particles under the trade name of "Percoll" (Pharmacia). The development has been described in several review articles (refs. 6 - 8).

Silica Sols

The colloidal silica solution we used (Ludox HS, E.I. duPont de Nemours) is colourless with a density of 1.3 g/ml and a viscosity of less than 5 centipoise. The colloidal particles have a mean diameter of about 20 nm but show a great polydispersity (Fig. 1). When a homogeneous solution of the colloid is centrifuged at high speed, transient concentration gradients - and density gradients - will be formed due to the polydispersity and the different sedimentation rates of the particles. In the presence of a polymer like dextran such gradients can be quite linear as seen in Fig. 2 (ref.9).

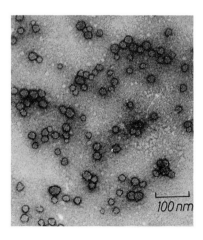

Fig. 1. Electromicroscopy of colloidal silica particles (Ludox HS) contrasted with 1% uranyl acetate.

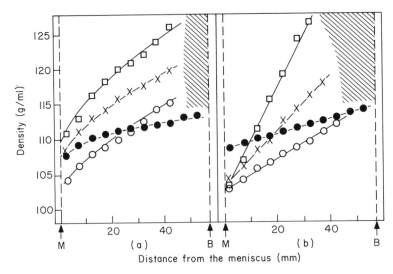

Fig. 2. Density gradients generated in colloidal silica (Ludox HS) in
the presence of dextran (mol.wt. 2x10^6) during centrifugation for
(a) 30 min and (b) 60 min at 35,000 rev./min. (O) 15% Ludox HS and 2%
dextran; (x) 24% Ludox HS and 2% dextran; (□) 30% Ludox HS and 2%
dextran; and (●) 15% Ludox HS and 10% dextran. (From ref.9. Reproduced
by kind permission of the publisher.)

Percoll

A pure silica sol is toxic to cells and causes hemolysis of red blood cells. However, the
dextran used in the early experiments protected the cells and this lead to a search for a
polymer that was non-toxic and that adsorbed firmly to the surface of the silica particles.
The final choice was polyvinylpyrrolidone which can be used effectively to coat the colloid
(ref.10). The average dimensions of a Percoll particle are given in Fig. 3 (refs.11-14).
    The coating of the silica with polyvinylpyrrolidone endowed the colloid with several
useful properties. Except for being non-toxic it also became more stable at physiological
salt concentration and pH (ref.12). Furthermore, the osmolarity of the colloid decreased
(ref.12) and the colloidal particles did not adsorb to the cell surfaces. In a
sedimentation equilibrium experiment we were able to reach a colloid concentration of 0.58
g/ml in physiological salt solution, which means that the colloid was stable even when the
particles were near maximal packing (ref.13). The extreme stability of the colloid even at
high salt concentrations was demonstrated by Price et al. (ref.15) when separating
microorganisms from ocean water.

PERCOLL®

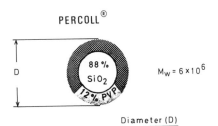

$M_W = 6 \times 10^6$

	Diameter (D)
Dry particle	21-22 nm
Hydrated particle (H$_2$O)	35 nm
— " — (0.15M NaCl)	28-30 nm

The PVP layer is 0.6 nm thick

Fig. 3. Physical parameters of an average Percoll
particle. (From ref.14. Reproduced by kind
permission of the publisher.)

FORMATION OF PERCOLL GRADIENTS

As described above gradients of Percoll can be prepared in situ by centrifugation at high
speed but also by the use of conventional mixing chambers. In 1979 Haff introduced another
gentle technique to form gradients of Percoll by freezing and thawing the solution (ref.
16). In cell biological work the gradients should be isotonic throughout and the problem
with uneven salt distribution by freezing and thawing was solved in 1984 (ref.17).

Various techniques were initially used to record the density gradients formed in Percoll (ref.8). A simple technique was introduced by Kågedal and Pertoft (ref.18) who stained gel beads of different densities with different dyes. When these beads are banded isopycnically the density gradient can be read from the banding pattern (Fig. 4).

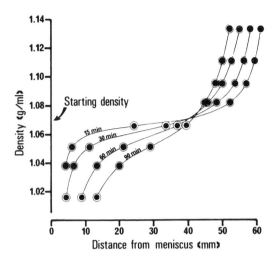

Fig. 4. Development of gradients of Percoll during centrifugation in an angle-head rotor, 8x14 ml (MSE Superspeed Centrifuge); starting density 1.07 g/ml; solvent 0.15 M NaCl. Running conditions: 20,000 g_{av} for 15, 30, 60 and 90 min. The density gradient was monitored by means of coloured "Density Marker Beads" (From ref.19. Reproduced by kind permission of the publisher.)

SEPARATION OF CELLS AND TISSUES

Percoll can be used both for rate sedimentation and isopycnic banding of cells. The principle of separating cells according to sedimentation velocity is old, e.g. red and white blood cells separate if blood is allowed to stand. A more advanced technique to separate cells according to sedimentation rate was described by Lindahl (ref.20) who constructed the counter-streaming centrifuge; the principle was reintroduced twenty years later and named "elutriation" (ref.21). When Percoll gradients have been used for separation according to sedimentation velocity, the experiments have usually been performed at unit gravity (ref. 22) or very low g-values in a centrifuge (ref.23), but Percoll has also been used in combination with elutriation rotors (ref.24).

Most cell separations in Percoll are based on isopycnic banding. There are now several hundred applications in which cells from animal tissues or blood have been fractionated according to density (ref.8). The density at which the cells band in an isotonic Percoll gradient corresponds to the physiological density and is usually much lower than the density at which the cells band in other density media, e.g. sucrose. The cells remain viable after banding in Percoll, even after centrifugation at very high g-values, and can be used for studies of intact cell functions including growth in cell culture.

Examples of problems which have been solved by means of Percoll centrifugations are the purification of human X-bearing sperms (ref.25) (the problem for which Lindahl constructed his counter-streaming centrifuge); separation of cytoplasts (enucleated cells) from intact cells (ref.26); synchronization of cell growth in cultures of bacteria and yeast (ref.27), malaria parasites (ref.28) and eucharyotic cells (ref.29, 30); separation of bacteria in waste water (ref.31); fractionation of plankton (ref.15); isolation of microfilariae from blood (ref.32, 33); and production of pure cultures of individual cells in liver (ref.34).

The use of Percoll is not confined to cells but can also be used for pieces of tissues or whole organs. A simple technique to isolate functionally intact pancreatic islets by sedimentation in Percoll was described by Buitrago et al. (ref.35) and the final purification of the various types of islet cells was achieved by a combination of Percoll centrifugation and elutriation (ref.36). Akerström et al. determined the isopycnic density of human parathyroid glands in Percoll gradients and could thereby determine the ratio of gland cells to fat cells in the tissue (ref.37). Tengvar et al. eqn (ref.38) were able to diagnose brain oedema by the same technique.

A recent application of Percoll is as a stabilizing medium in electrophoresis of cells (ref.39).

SEPARATION OF SUBCELLULAR PARTICLES

Cell organelles such as nuclei, mitochondria, lysosomes, peroxisomes and membrane vesicles can be purified by isopycnic centrifugation in Percoll. A prerequisite for banding is that the particles have sedimentation rates well in excess of that of the colloidal silica. At present the technique therefore is unsuitable for viruses smaller than poliovirus (110 S).

The densities of subcellular particles in Percoll and sucrose gradients are recorded in Fig. 5. Like intact cells the subcellular particles exhibit lower densities in Percoll than in other media due to the isotonicity of the gradients. The importance of the osmolarity of the surrounding medium was clearly demonstrated by Lagercrantz et al. (ref.40) for catecholamine storing granules. The authors varied the sucrose concentration in a Percoll gradient and showed an increased granule density with increasing concentration.

The density of a subcellular particle can also vary physiologically. This was demonstrated for the first time by Pertoft et al. (ref.41) who followed the catabolism of protein in liver lysosomes. The lysosome became denser during the process. Such a study could not have been carried out in a hypertonic density gradient.

An interesting observation was made when polystyrene particles of different sizes but with the same absolute density were banded in colloidal silica (ref.9). The particles were banded according to size, the smallest having the lowest isopycnic density in the gradient. Similarly when some viruses were studied the very asymmetric tobacco virus showed an extremely low isopycnic density (ref.42). The phenomenon was explained when the density was measured in silica colloids of different particle diameters. The larger the colloidal particle the lower the isopycnic density of the virus. The colloidal particles are excluded from a volume around the virus (Fig. 6) leaving a "hydration shell" which is determined by the surface area of the virus and the radius of the colloidal particle. The larger the relative size of the hydration shell the lower the isopycnic density.

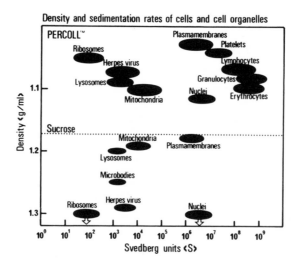

Fig. 5. Approximate sedimentation rates and isopycnic banding densities of herpes virus, particles in a rat liver homogenate and human blood cells in gradients of Percoll and sucrose. (From ref.19. Reproduced by kind permission of the publisher.)

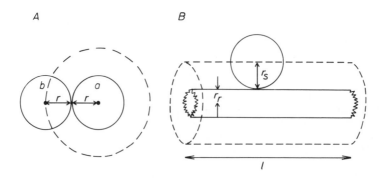

Fig. 6. Demonstration of the exclusion of spherical Percoll particles from the domains of (A) a spherical and (B) a rod-like molecule (e.g. tobacco mosaic virus). The centre of the Percoll particle can never come closer to the surfaces of the molecules than its own radius, thus leaving a "hydration shell" around the molecules (broken lines). It is the over-all density of the "hydrated" molecule that determines where it bands in a Percoll gradient.

CONCLUSIONS

The ultracentrifuge has been used for both analytical and preparative work on macromolecules. Since the time of The Svedberg the focus of interest has moved from macromolecules to more complex biological structures such as cells and cell organelles. Consequently there has been a rising need for methods to purify and analyse such structures. Coated colloidal silica media represent a significant advance in the centrifugation of labile biological materials as it can be used to form isoosmotic density gradients. The new medium has been used to reveal subpopulations of both cells and cell organelles not previously expected. The introduction of silica solutions has provided a material, which not only gives a true density but also quantifies the absolute density range and distribution. There is, however, still a great need for new and refined methods to isolate and analyse biological particles.

The Svedberg´s work on inorganic colloids and the ultracentrifuge has been the basis for still another technique utilized in biology and medicine.

Work reviewed in this paper has been supported by the Swedish Medical Research Council (project 3X-4).

REFERENCES

1. T.C. Laurent, Fed. Proceed. **25**, 1128-1134 (1966).
2. M.K. Brakke, J. Am. Chem. Soc. **73**, 1847-1848 (1951).
3. E.G. Pickles, Methods Med. Res. **5**, 107-113 (1952).
4. M. Meselson, F.W. Stahl and J. Vinograd. Proc. Natl. Acad. Sci USA **43**, 581-588 (1957).
5. C. de Duve, J. Berthet and H. Beaufay, Progr. Biochem. Biophys. Chem. **9**, 325-369 (1959).
6. H. Pertoft and T.C. Laurent. In Modern Separation Methods of Macromolecules and Particles (T. Gerritsen, ed.), pp. 71-90. Wiley (Interscience), New York (1969).
7. H. Pertoft and T.C. Laurent. In Methods of Cell Separation (N. Catsimpoolas, ed.), pp. 25-65. Plenum, New York (1977).
8. H. Pertoft and T.C. Laurent. In Cell Separation: Methods and Selected Applications (T.G. Pretlow and T.P. Pretlow, eds.), Vol 1, pp. 115-152. Academic Press, New York (1982).
9. H. Pertoft, Biochim. Biophys. Acta **126**, 594-496 (1966).
10. H. Pertoft, K. Rubin, L. Kjellén, T.C. Laurent and B. Klingeborn, Exp. Cell Res. **110**, 449-457 (1977).
11. H. Pertoft, T.C. Laurent, T. Laas and L. Kagedal, Anal. Biochem. usual **88**, 271-282 (1978).
12. T.C. Laurent, H. Pertoft and O. Nordli, J. Colloid Interface Sci. eq **76**, 124-132 (1980).
13. T.C. Laurent, A.G. Ogston, H. Pertoft and B. Carlsson, J. Colloid Interface Sci. eq **76**, 133-141 (1980).
14. H. Pertoft and T.C. Laurent. In Cell Function and Differentiation (G. Akoyunoglou et al., eds.) Part A, pp. 95-104. Alan R Liss, New York (1982).
15. C.A. Price, E.M. Reardon and R.R.L. Guillard, Limnol. Oceanogr. eq **23**, 548-553 (1978).
16. L.A. Haff, Prep. Biochem. eq **9**, 149-156 (1979).
17. A.J. Cooper, J.A. Smallwood and R.A. Morgan, J. Immunol. Methods eq **71**, 259-264 (1984).
18. L. Kagedal and H. Pertoft, Abstr. 3341, 12th FEBS-Meeting, Dresden (1978).
19. H. Pertoft, T.C. Laurent, R. Seljelid, G. Akerström, L. Kagedal and M. Hirtenstein. In Separation of Cells and Subcellular Elements (H. Peeters, ed.), pp. 67-72. Pergamon Press, Oxford (1979).
20. P.E. Lindahl, Nature eq **161**, 648-649 (1948).
21. C.R. McEwen, R.W. Stallard and E.The. Juhos, Anal Biochem. eq **23**, 369-377 (1968).
22. T. Nethanel, R. Kinsky, N. Moav, R. Brown, M. Ran and I.P. Witz, J. Immunol. Methods **41**, 43-56 (1981).
23. A. Tulp, M.W. Kooi, J.B.A. Kipp, M.G. Barnhoorn and F. Polak, Anal. Biochem. eq **117**, 354-365 (1981).
24. C.G. Figdor, W.S. Bont, I. Touw, J. de Roos, E.E. Roosnek and J.E. de Vries, Blood eq **60**, 46-53 (1981),
25. S. Kaneko, S. Oshio, T. Kobayashi, H. Mohri and R. Ilzuka, Biomed. Res. eq 5, 187-194 (1984)
26. W. Bossart, H. Loeffler and K. Bienz, Exp. Cell Res. eq **96**, 360-366 (1975).
27. R.D. Dwek, L.H. Kobrin, N. Grossman and E.Z. Ron, J. Bacteriol. **144**, 17-21 (1980).
28. E.M. Rivadeneira, M. Wasserman and C.T. Espinal, J. Protozool. eq **30**, 367-370 (1983).
29. D.A. Wolff and H. Pertoft, J. Cell Biol. eq **55**, 579-585 (1972).
30. M.R. Loken and H.E. Kubitschek, J. Cell. Physiol. eq **118**, 22-26 (1984).
31. P. Scherer, J. Appl. Bacterol. eq **55**, 481-486 (1983).
32. H. Feldmeier, U. Bienzle and D. Schuh, Trans. Roy Soc. Tropical Med. Hygiene eq **75**, 251-253 (1981).
33. D.J. Grab and J.J. Bwayo, Acta Tropica eq **39**, 363-366 (1982).
34. B. Smedsrød, S. Eriksson, J.R.E. Fraser, T.C. Laurent and H. Pertoft. eq In Sinusoidal Liver Cells eq (D.L. Knook and E. Wisse, eds.), pp. 263-270. eq Elsevier/North-Holland Biomedical Press, Amsterdam (1982).

35. A. Buitrago, E. Gylfe, C. Henriksson and H. Pertoft, <u>Biochem Biophys. Res. Commun.</u> eq
     79, 823-828 (1977).
36. D.G. Pipeleers and M.A. Pipeleers-Marichal, <u>Diabetologica</u> eq **20**, 654-663 (1981).
37. G. Åkerström, H. Pertoft, L. Grimelius and H. Johansson, <u>Acta Path. Microbiol. Scand.</u>
     <u>Sect. A.</u> eq **87**, 91-96 (1979).
38. Ch. Tengvar, M. Forssén, D. Hultström, Y. Olsson, H. Pertoft and A. Petterson,
     <u>Acta Neuropathol. (Berl.)</u> eq **57**, 143-150 (1982).
39. D.N. Ku, <u>Sep. Purif. Methods</u> eq **11**, 71-82 (1982).
40. H. Lagercrantz, H. Pertoft and L. Stjärne, <u>Acta Physiol. Scand.</u> eq **78**, 561-566 (1970).
41. H. Pertoft, B. Wärmegård and M. Höök, <u>Biochem. J.</u> eq **174**, 309-317 (1978).
42. H. Pertoft, L. Philipson, P. Oxelfelt and S. Höglund, <u>Virology</u> eq **33**, 185-196 (1967).

IV Sedimentation Analysis of Macromolecules

13. Analytical Sedimentation – An Overview of Experimental Techniques and Theoretical Developments

Lars-Olof Sundelöf

Institute of Inorganic and Physical Chemistry, Faculty of Pharmacy, Biomedical Centre, Uppsala University, Uppsala, Sweden

Abstract – A general description is given of problems in colloidal and macromolecular chemistry amenable to ultracentrifugational investigation. The most important experimental aspects are described: Field requirements and cell dimensions, drive principles, rotors, cells, recording and sensitivity aspects. The physics of ultracentrifugation is described in outline. The main features of the transient(velocity) case and of the equilibrium case are treated. Finally a few remarks are given as to density gradient sedimentation and chemically interacting systems.

INTRODUCTION

During the last century it was already a well established fact that there exists some kind of relationship between rate of sedimentation and size of macroscopic particles in a liquid dispersion acted upon by the gravitational field. If the particles are small enough, as is the case for many colloids, the sedimentation rate in the gravitational field becomes too slow to be observed. The rate of sedimentation can be magnified, however, by spinning the dispersion in a centrifuge. This was early realized by The Svedberg and he adopted this principle to create a tool whereby the size and size distribution of colloidal particles could be determined. That was the beginning of the development of a very powerful technique nowadays known as analytical sedimentation and applicable even to the study of small molecules.

Here only a brief overview of this vast subject will be given and mainly from the point of view of physical macromolecular chemistry.

SOME GENERAL MACROMOLECULAR/COLLOIDAL PROBLEMS

A central parameter for almost any physico-chemical discussion of macromolecular systems is the molecular mass of the molecules involved. Hence precise methods for molecular weight determination are needed. Just as central, however, is a knowledge of how the segments building up the molecules are distributed in space whether in the form of dense spherical or ellipsoidical particles, statistical coil structures or in a more or less extended rodlike form. The molecular mass in combination with its spatial distribution tells us about the average volume over which the molecule can be active in solution. Since the centrifugal force is proportional to the mass and since the frictional resistance to flow depends on the average geometrical form of the segment distribution, it is intuitively clear that sedimentation can provide the desired information.

Almost always a macromolecular sample is nonhomogeneous in the sense that there is a distribution of "properties" of some sort. The chemical composition can be the same but the molecular mass can vary, i.e. there is a molecular weight or size distribution. The sample can contain different species with different physical properties: mass, conformation, partial specific volume. For derivatives, the degree of substitution as well as the distribution of substituents along the chain may differ from sample to sample again affecting the physical properties determining the rate of sedimentation. Branching and tacticity are other features that affect especially the spatial segment distribution sufficiently to give the sample new properties. Clearly such distributions of properties will be reflected in the rate of sedimentation and can thus be studied experimentally.

When concentration increases, molecular contacts become more probable and interactions of various types begin to show up. Such interactions can be thermodynamic in nature deriving from "contacts" between the segment clouds of two or more neighbouring molecules affecting the size and hence the hydrodynamical properties of the sedimenting molecules. Even non deformable particles begin to interact hydrodynamically at elevated concentrations. If, finally the molecules are able to react chemically – they may for instance associate – this will affect the mass and possibly also the geometrical size. Evidently interactions of the type indicated could be observed through the rate of sedimentation.

From these few remarks it is obvious that analytical sedimentation is a tool of fairly broad applicability to many general problems in macromolecular chemistry.

EXPERIMENTAL ASPECTS

The smaller the particles the stronger must the accelerating field be to achieve a measurable sedimentation. Fairly simple calculations show that to achieve a sedimentation rate of about 1 cm per hour for a particle with molecular mass 100,000, a speed of revolution of 1,000 r.p.s. will be needed at a radial distance of some 10 cm. These realistic average figures do provide the main guidelines for the design of an appropriate experimental setup of an ultracentrifuge with regard to

- physical size of rotor
- accelerating field requirement
- size of cell for monitoring the sedimentation process.

Drive principles

In the earlier development of analytical sedimentation the oil turbine technique played a dominant role. Here the rotor is supported on both sides of the main body by bearings and is driven by turbine wheels at both ends of the axis. When properly adjusted this provides very good running performance but the balancing requirements are very strict. The rotors are heavy and run in big metal housings allowing only a restricted temperature range. The optics is usually straight which is an advantage.

In later developments the electrical drive principle has been generally adopted usually in combination with a vertical, flexible shaft axis. This design has the advantage that the spinning rotor is self-adjusting once it is statically balanced. The rotor can be made much less heavy, it spins in a vacuum chamber where the temperature can be regulated over a wide temperature range. The electrical drive makes feedback speed control feasible and by electronic multiplexing multihole rotors can be used to speed up experiments. The optic axis is folded to make the design more compact.

In some cases magnetically suspended rotors have been used. This design has been advanced due to the need in some long run equilibrium experiments to avoid vibrations. Some designs have even used air cushions for this purpose. The magnetic suspension technique, however, requires quite elaborate control equipment and would be no routine method.

Rotors

The design of rotors is critical to the success of sedimentation experiments. On one hand they must be strong enough to withstand the centrifugal field (normally up to some 300,000 g), on the other hand they should be possible to handle manually. The development has thus gone from special steel to light weight alloys and most recently titanium to allow high speed runs without too much deformation.

During acceleration the rotor is stretched leading to an increase in the radial distance of the cell position of about half a percent. Likewise during acceleration the rotor is adiabatically cooled, the temperature drop being of the order of $1^{\circ}C$. Such a temperature drop must be carefully observed in low concentration runs where convections are likely to disturb the results. Closely connected with these aspects is the problem of thermal stability. It is especially important that no heat leaks from the centre of rotation in the radial direction, since that might cause density inversions and convections.

Vibrations must of course be avoided. In high speed runs they usually do not cause much of a problem in part due to the background density gradient in the solvent. Equilibrium runs are much more sensitive to such disturbances, however, and at low speeds the possibility of resonance frequencies must be carefully watched.

Cells

The cells are sector shaped to allow radial sedimentation. The sector angle varies from 2° to 4° and proper alignment with respect to the centre of rotation is mandatory. Modern cells are of a sandwich design: window, centrepiece, window - all placed in proper holders and pushed into a cylindrical mantle and locked by a nut. Cells should be easy to fill and close and if refractive index recording is used the windows should show no strain. For analytical sensitivity reasons the centrepiece thickness varies up to about 30 mm. In multiple hole rotors wedged windows are sometimes used to distribute the records side by side ("optical multiplexing"). For interference recording, double sector cells must be used requiring very precise manufacturing of the centrepiece with respect to flatness and non-deformability.

To create solution/solvent boundaries in the cells during a run special synthetic boundary cells have been developed allowing very slow sedimentation to be observed and also the diffusion spreading of the interface to be monitored. From such experiments the diffusion coefficient can be calculated and combined with the sedimentation coefficient to give the molecular weight according to the Svedberg formula. This makes a separate diffusion equipment unneccessary for routine measurements. Other special cells, for instance for separation of components directly in the analytical run, have been developed.

Good cells are often the key to successful experiments and modern materials and refined mechanical designs have greatly advanced the possibilities of utilizing the sedimentation technique.

Recording and sensitivity requirements

The most general and also the most precise method of recording concentration profiles is through the use of refractive index. This can be done in two principally different ways. One is based upon the deviation of a light beam when passing through a concentration gradient, the angular deviation being proportional to the thickness of the solution along the optical axis and to the refractive index gradient in the radial direction of the cell. The first method based on this principle was the Lamm scale method and later the schlieren method was developed and now exists in a variety of forms. In the schlieren method light itself is allowed - by proper astigmatic optics - to draw the gradient curve, which is the most direct way of observing the position and shape of the sedimenting boundary.

The other method based on refractive index is the interference technique, where two separate coherent light beams pass the measuring and the reference channels of the cell, respectively, before being brought together optically to create a fringe pattern super-imposed on the radial coordinate. The fringe displacement thus gives the concentration at the corresponding position. When properly used this principle is very precise.

In later years the light absorption method has been taken up again as a very sensitive means of detecting low concentrations. Modern electronics and light sensing devices also make it very precise.

A great number of optical systems and design principles have been advanced for the methods mentioned above and in general one can say that the sensitivity increases from the deviation methods ($\Delta c \approx 0.1$ g/dl), through the interference methods ($\Delta c \approx 0.01$ g/dl) to the light absorption method ($\Delta c \approx 0.001$ g/dl), the figures of course being only rough indications of the concentration step across the boundary.

The sensitivity requirements may need a few comments. If only the sedimentation coefficient is of interest it is sufficient to record the position of the sedimenting boundary as a function of time. For that a fairly simple recording system will be sufficient. The situation is quite different if the shape of the boundary is needed, which is the case if for instance distribution properties are studied, as mentioned above.

THE PHYSICS OF ULTRACENTRIFUGATION

The physical situation of a particle in an accelerating centrifugal field, $\underline{a} = \omega^2 \underline{r}$ (where ω is the angular velocity and $\underline{r}$ is the radial distance), is characterized by the particle mass, $\underline{m}$, the partial specific volume, $\overline{v}$, and the density of the surrounding medium, $\cdot$. The kinetic energy transferred to the particle through the accelerating field is dissipated in the medium as frictional resistance giving the particle a constant velocity $\underline{v}$. From the force balance one gets

$$\underline{v} = \underline{s} \cdot \underline{a} \tag{1}$$

where $\underline{a}$ is the "disturbing" field and $\underline{v}$ is the particle velocity "response". The proportionality factor $\underline{s}$, the sedimentation coefficient, represents the sedimentation velocity for unit field and is given by

$$\underline{s} = \frac{m \ (1 - \overline{v}\rho)}{f} \tag{2a}$$

where f is the frictional coefficient which, apart from medium properties, depends both on molecular mass and conformation. Eqn (2a) gives a relationship between observed rate of sedimentation and molecular properties, i.e.

$$\underline{s} \propto \frac{M}{f} \tag{2b}$$

which is fundamental to all sedimentation analysis.

The accelerating field induces a sedimentation flow, $\underline{J}_{sed} = \underline{v} \cdot c$, leading to a redistribution of concentration, $c(\underline{r}, t)$, and the concentration gradient thus created sets up a restoring diffusion flow, $\underline{J}_{diff} = -\frac{\partial c}{\partial \underline{r}}$, the total flow being given by

$$\underline{J}_{tot} = \underline{J}_{sed} + \underline{J}_{diff} \quad \cdot \tag{3}$$

Rewritten as a differential "continuity" equation this becomes

$$\frac{\partial c}{\partial \underline{t}} = \frac{1}{\underline{r}} \frac{\partial}{\partial \underline{r}} \ [(D \ \frac{\partial c}{\partial \underline{r}} - \omega \underline{rsc})\underline{r}] \quad \cdot \tag{4}$$

Eqn (4) can be solved for various initial conditions and boundary values. The time evolution of the sedimentation process can thus be shown to proceed from the homogeneous initial solution over the transient ("velocity") stage to the final equilibrium (time independent) state.

It could be mentioned that for a rigorous treatment of multicomponent systems the formalism of irreversible thermodynamics is of great advantage.

Some observations on the transient (velocity) case

Obviously the boundary motion will be essentially exponential in time according to the relation

$$\underline{r} \propto e^{\underline{s}\omega^2 \underline{t}} \quad . \tag{5}$$

The sector shape of cells leads to the "square law" dilution phenomenon, i.e. the solution under the boundary is continuously diluted during the experiment. For most macromolecules in solution the concentration dependence of $\underline{s}$ is very pronounced in dilute solution, which requires extreme precaution in extrapolating the data to infinite dilution where the molecules behave as individual units. Frequently this concentration dependence is hyperbolic

$$\underline{s} = \frac{\underline{s}_o}{1 + \underline{k} \cdot \underline{c}} \quad . \tag{6}$$

The solvent compressibility becomes noticeable at higher speeds and especially in organic solvents. Apart from the density this affects the partial specific volume and in particular the viscosity. The pressure variation from the meniscus to the cell bottom often varies from 1 to 200 atm which means a variation in the sedimentation coefficient of the order of 10 to 20%. From the practical point of view the pressure dependence can be taken as linear

$$\underline{s}_p = \underline{s}_o (1 - \alpha \underline{P}) \quad . \tag{7}$$

In order to apply Svedberg´s formula for molecular weight determination

$$\langle \underline{M} \rangle = \frac{\langle \underline{s} \rangle_{oo} RT}{\langle \underline{D} \rangle_o (1 - \overline{\underline{v}}\rho)} \tag{8}$$

the sedimentation coefficient must not only be extrapolated to infinite dilution but also to "zero" pressure. Reliable routine procedures for this are now available.

As seen from eqn (2b) the frictional conditions play a decisive role for the rate of sedimentation and a number of more or less advanced theories have been put forward. A general feature is, however, that the frictional coefficient (at infinite dilution) becomes proportional to a characteristic length in the molecule, $\langle \underline{L} \rangle$, and to its shape

$$f_o \propto \langle \underline{L} \rangle \, \Psi_{shape} \quad . \tag{9}$$

If the shape factor is known - or does not vary - the sedimentation analysis technique can thus be used as a sensitive tool for conformational studies and particularly so for particles with a less extended segment distribution. In this way it serves as a complement to viscosity studies.

Everything discussed so far refers only to the motion of the boundary itself. But the analysis can naturally be brought further into a detailed investigation of the boundary shape. Then a number of factors must be taken into account and mathematics becomes involved; frequently closed solutions cannot be obtained. In recent years computer solutions have been tried.

A special complication is the mutual concentration dependence of all species present in the boundary region, the so called Johnston-Ogston effect, and which must be considered in determinations of distributions of sedimentation coefficients.

A second complication is the diffusion broadening which tends to "smear out" the size distribution information present in the boundary. Since the separation of species is proportional to the time, $\underline{t}$, whereas the diffusion broadening increases only with $\sqrt{\underline{t}}$, proper extrapolation to infinite time can free the data of this effect.

Some observations on the equilibrium case

In equilibrium sedimentation all information is extracted from the detailed shape of the concentration distribution in the cell. Although the complex hydrodynamic interaction is thus eliminated, the thermodynamic interaction between species in the continuously varying concentration created by superimposed distributions tends to make the analysis less straightforward. However, molecular weight averages of higher order can be obtained and with some precision also virial coefficients. Attempts to resolve the size distribution by convolution methods have not been very successful due to the finite length of the solution column in the radial direction which leads to truncation errors.

An important experimental aspect is the necessity to establish a well defined reference point for the absolute concentration calculation. This has lead to the development of the high speed equilibrium method where the meniscus becomes depleted of substance and thus can serve as the desired reference.

Sometimes the time to attain equilibrium becomes prohibitively long and a compromise must be made between the precision obtained and the time of the experiment. To some degree a "speed programme" can shorten this time.

Closely related to the true equilibrium is the "approach to equilibrium situation". Here one utilizes the fact that at the meniscus and at the bottom the total flow must be zero and by measuring the concentration and the concentration gradient simultaneously at these points an expression giving the molecular weight directly can be used, e.g.

$$\frac{1}{\omega^2 r_m} \left(\frac{1}{c} \cdot \frac{dc}{dr} \right)_m = \frac{s}{D} \propto M \qquad . \tag{10}$$

Density gradient sedimentation

In order to separate species of different chemical composition it is sometimes possible to resort to variations in the partial specific volume. By allowing the sample to sediment in a background density gradient the different species will distribute themselves in bands according to "matched density". This has in certain studies provided a powerful means of composition analysis. Even solvation properties can be studied. It is also possible to analyse the band shape for more detailed information.

Chemical interaction

Obviously, if the molecules present are able to react, new molecular masses will be formed which tend to redistribute in the sedimentation cell. Inherent in the sedimentation analysis lies therefore the possibility of studying chemical reactions, in particular dissociations and associations. This must be done, however, in the light of the principles discussed in this presentation and it can be firmly stated that the general problem is not simple. Often additional information is at hand, however, and thus the sedimentation analysis may provide conclusive arguments.

ACKNOWLEDGEMENT

It has been a privilege to have been asked to give this presentation at the Svedberg memorial symposium and for this I express my thanks to the organizers. This is a review of a very broad field emanating from Uppsala University and the facts put forward have been collected from many sources and not least through stimulating personal contacts. For this I would like to thank three friends who have meant very much to me over the years, namely, Professors Stig Claesson, Kai Pedersen and Jack Williams.

REFERENCES

No detailed references will be given since the presentation is in overview form. The reader is referred to the excellent monographs on the subject, especially

T. Svedberg, K.O. Pedersen, The Ultracentrifuge. eq Clarendon Press, Oxford (1940).
H.K. Schachman, Ultracentrifugation in Biochemistry. eq Academic Press, New York (1959).
H. Fujita, Mathematical Theory of Sedimentation Analysis. eq Academic Press, New York (1962).
Ultracentrifugal Analysis in Theory and Experiment. J.W. Williams (ed.). Academic Press, New York (1963).

14. The Ultracentrifuge – A Source of Information on Macromolecules

A.J. Staverman

Gorlaeus Laboratories, P.O. Box 9502, 2300 RA Leiden, The Netherlands

Abstract – The significance of the ultracentrifuge for the determination of molecular weights and molecular weight distribution (MWD) has declined since the advent of reliable alternative methods.

Unique information about intermolecular interaction can be obtained from measurements of the effect of the concentration of macromolecules on the sedimentation velocity of these or other macromolecules.

In order to describe results of a large number of such experiments performed by Peeters, classical models of the equivalent sphere and of congruent bodies for a series of homologous macromolecules at $T=\theta$ are definitely inadequate. Also the model of Pyun and Fixman fails to describe the experimental results. An alternative model for the intermolecular interaction, containing the basic idea of Pyun and Fixman, is sketched and describes the experiments correctly in a qualitative way. Systematic measurements of sedimentation velocities in binary and ternary solutions of macromolecules may allow discrimination between at least four different interaction effects: backflow, field distortion, intertwinement and collisions.

DETERMINATION OF MWD (MOLECULAR WEIGHT DISTRIBUTION)

In their first experiments on proteins Svedberg and Pedersen were lucky and clearly surprised to find that these polymers are mono- or pauci-disperse. Soon it was realized that this favourable property of natural polymers is not shared by synthetic polymers. Since in the old days uncertainty about the MWD disturbed many investigations on synthetic polymers, many efforts have been made to determine MWD´s.

This problem is solved by the development of reliable GPC measurements, while sedimentation equilibrium measurements can supply additional information. However, from a careful analysis of $c(r,t)$-plots in sedimentation velocity experiments much more information can be derived than a single average sedimentation constant and many authors have tried to convert this into information about the MWD. Svedberg in his early papers (refs. 1, 2) always mentioned "Determination of Size and Distribution of Size..." and Fujita devotes considerable attention to the determination of MWD in his book (ref. 3). Baldwins method of moments appeared most promising (ref. 4). Baldwin has shown that average molecular weights can be determined from measurements of moments of dc/dr over r, defined by

$$I_r^n(t) \equiv \int_{r_0}^{r_m} r^n \, (dc/dr) \, dr \quad .$$

$$(1)$$

Mulderije (refs. 5, 6) has suggested that measurements of moments over t, $I_t^n(r)$, as a function of r may possibly be experimentally attractive and allow the determination of interesting M-averages. If we define

$$I_t^n(r) \equiv \int_0^\infty t^n c(r,t) \, dt$$

$$(2)$$

or

$$J_t^n(r) \equiv \int_0^\infty t^n (\delta c/\delta r) \, dt$$

$$(3)$$

then I_t^n and J_t^n are measureable quantities. The exponent $n > 0$, need not be an integer.

Mulderije has shown, that a simple relation exists between average sedimentation coefficients $\langle \underline{s}^{-n} \rangle$ and $\underline{I}_t^{n-1}$.

By definition

$$\langle \underline{s}^{-n} \rangle \equiv \Sigma_i \underline{c}_i^0 \underline{s}_i^{-n} / \Sigma_i \underline{c}_i^0 \quad . \tag{4}$$

Mulderije has shown, that

$$\underline{I}_t^{n-1}/\underline{c}^0 = (2\omega^2)^{-n} \underline{K}(\underline{n},\underline{r}) \langle \underline{s}^{-n} \rangle \tag{5}$$

where the coefficient $\underline{K}(\underline{n},\underline{r})$ is a known function of $\underline{n}$ and $\underline{r}$. If the relation between $\underline{s}$ and $\underline{M}$ is known

$$\underline{s} = \underline{K}_s \underline{M}^b \tag{6}$$

we have

$$\langle \underline{s}^{-n} \rangle = \underline{K}_s^{-n} \langle \underline{M}^{-bn} \rangle \quad . \tag{7}$$

Thus average of negative powers of $\underline{M}$ can be determined. This is interesting if information about the low molecular tail of the MWD is desired. By choosing $\underline{n} = 1/\underline{b}$, the number average molecular weight

$$\underline{M}_n \equiv \langle \underline{M}^{-1} \rangle^{-1} \tag{8}$$

is determined. If a large value of $\underline{n}$ is chosen, large values of $\underline{t}$ count heavily in the moment integral and, therefore, slow small molecules contribute strongly.

The determination of the MWD from moments of time appears simpler than that from moments of $\underline{r}$, both experimentally and mathematically. If diffusion is negligible, every component contributes to the increment of $\underline{I}_t$ at one characteristic moment. Thus integrals over intervals of time correspond directly with parts of the MWD. However, one fundamental problem underlies, and in fact undermines, the reliability of interpretation of all sedimentation transport measurements of polydisperse polymers. While the effects of diffusion, pressure and concentration can be accounted for with correction terms or extrapolation procedures, complete ignorance, theoretical as well as experimental, exists with respect to the interaction between molecules of different size and mass, moving with different velocity.

I quote from Fujita's book (ref. 3) p. 194: "At present it seems by no means easy to find a theoretical guidance for this purpose if one recalls that for ternary solutions there was a formidable difficulty in the theory of coupled sedimentation".

Here Fujita referred to the above mentioned problem of interaction between molecules of different velocity and size. Knowledge about this problem is not only essential for the interpretation of sedimentation profiles of polydisperse polymers, but even more for the understanding of interactions and intertwining of macromolecules in general.

INTERMOLECULAR INTERACTIONS IN SEDIMENTATION TRANSPORT

The friction coefficient of a macromolecule, defined by

$$\underline{f} \equiv \frac{RT}{\underline{D}} \tag{9}$$

is an interesting quantity because its value depends on the structure of the macromolecule and its thermodynamic and hydrodynamic interaction with solvent and eventually other molecules. It follows from Svedberg's relation

$$\underline{M} = \frac{RT\underline{s}}{\underline{D}(1-\bar{v}\rho)} \tag{10}$$

that determination of the sedimentation constant $\underline{s}$ means determination of the ratio $\underline{M}/\underline{f}$. Since $\underline{M}$ can be determined in other ways, the ultracentrifuge is an efficient instrument for determining friction coefficients and their dependence on boundary conditions.

Binary systems

For dilute solutions the dependence of the friction coefficient on the concentration can be written

$$\underline{f} = \underline{f}(0) (1 + \underline{k}_c \underline{c}) \tag{11}$$

with c in g per ml. Alternatively one can write

$$\underline{f} = \underline{f}(0) \ (1 + \underline{k}^{\phi} \ \phi) \tag{12}$$

where ϕ is the volume fraction and $\phi = \overline{v} \ c$ (13)

and $\overline{v}$ is the specific volume of the solute.

The expression (12) is useful since it permits the comparison of experimental values of $\underline{k}^{\phi}$ with theoretical values calculated for hard spheres, ellipsoids, rods and cylinders. The determination of $\underline{k}$-values for polymers gives information about the hydrodynamic interaction between the macromolecules provided that the thermodynamic interaction is known. In θ-solutions the thermodynamic interaction is such that the distribution of molecules is also very nearly random in semi-dilute solutions. There measurements at $\underline{T} = \theta$ are of particular interest.

The sedimentation coefficients, $\underline{s}$, and also the viscosity indices, $[\eta]$, of polystyrene in cyclohexane at $\underline{T} = \theta = 34°C$ have been measured by many investigators. Peeters (refs. 7, 8) repeated these measurements very carefully with sharp polystyrene fractions over a large range of molecular weights. The fractions were obtained from different sources. When the dispersity was too high, sharp fractions were extracted by laborious preparative gel permeation chromatography.

The measured values of $[\eta]$ and $\underline{s}^0$, the limiting value of $\underline{s}$ at low concentration, showed a strictly linear relation between these quantities and $\underline{M}^{\frac{1}{2}}$ over the entire range of molecular weights (Fig. 1) $10^4 \leqslant \underline{M} \leqslant 2.2 \times 10^6$.

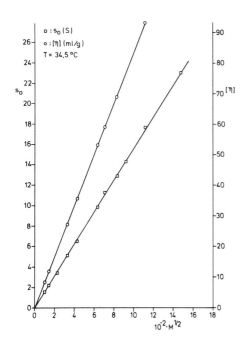

Fig.1. $\underline{s}^0$ and $[\eta]$ as functions of $\underline{M}^{\frac{1}{2}}$, (refs. 7, 8)

The proportionality is in accordance with the theory and justifies confidence. $\underline{M}$-values were measured from light scattering. The sedimentation measurements were performed with a Spinco E from Beckman Instruments.

For $[\eta]$ the best fit was obtained with

$$[\eta] = 8.26 \times 10^{-2} \ \underline{M}^{\frac{1}{2}} \ (ml/g) \tag{14}$$

and for $\underline{s}^0$:

$$\underline{s}^0 = 1.56 \times 10^{-2} \ \underline{M}^{\frac{1}{2}} \ (\text{Svedberg} = 10^{-3} \ s) \tag{15}$$

For a number of fractions Peeters measured sedimentation constants at increasing concentrations in order to determine $\underline{k}$-values. The results are shown in Fig. 2 which also shows results of earlier investigators. In contrast to all his predecessors Peeters found a non-linear relation between $\underline{k}$ and $\underline{s}$, which implies also a non-linear relation between $\underline{k}$ and $\underline{M}^{\frac{1}{2}}$.

Instead he summarizes his results in the following relation

$$\underline{k}^{\underline{c}} = 3.3 \ \underline{s}^0 + 0.07 \ (\underline{s}^0)^2 + \underline{v} \ ml/g \qquad (16)$$

where $\overline{v} = 0.934 \ ml/g$.

The theoretical implications of the non-linearity will be discussed later. The fact that earlier investigators did not find this phenomenon may be due to some polydispersity of their fractions and/or to an insufficient range of molecular weights. Especially for the fractions of large $\underline{M}$, polydispersity will lower the measured values of $\underline{s}$ (ref. 9), thus masking the upturn in Peeters' curve. In his thesis he discusses possible sources of error and shows that these cannot give rise to the deviation form linearity.

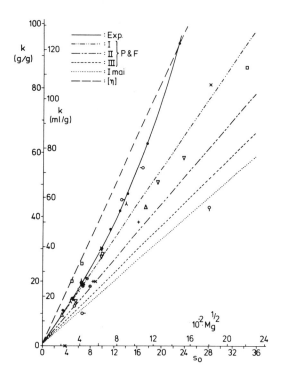

Fig. 2. The parameter $\underline{k}^{\underline{c}}$ as a function of $\underline{M}^{\frac{1}{2}}$, (refs. 7, 8).
...... and ———— are results of Peeters. I, II and III are the theories of Pyun and Fixman. Results of different authors are indicated.

Ternary Systems

Peeters also performed measurements of sedimentation transport on mixtures of two sharp polystyrene fractions in θ-solution (ref. 18). The sedimentation velocity of both components was measured and the effect of the presence of component $\underline{j}$ on the sedimentation velocity of component $\underline{i}$ was established. These effects reflect the interaction between macromolecules moving with different velocity. Besides the ultracentrifuge few instruments will allow the rapid determination of this interaction. Diffusion experiments may give similar information but presumably, such experiments will be very laborious. In ternary systems eqn (11) is generalized as follows.

$$\underline{f}_{\underline{i}}(\underline{c}_{\underline{i}}, \underline{c}_{\underline{j}}) = \underline{f}_{\underline{i}}(0) \ (1 + \Sigma_{\underline{j}} \underline{k}^{\underline{c}}_{\underline{i}\underline{j}} \underline{c}_{\underline{j}}) \qquad . \qquad (17)$$

The coefficient $\underline{k}_{\underline{i}\underline{j}}$ describes the effect, acceleration or retardation, of component $\underline{j}$ on component $\underline{i}$. The coefficients $\underline{k}_{\underline{i}\underline{j}}$ and $\underline{k}_{\underline{j}\underline{i}}$ are not related by Onsager relations. However, they are related to Klemms' friction coefficients, which do obey the O.R.R.. With eqn (17) the sedimentation coefficients are written:

$$\underline{s}^0_{\underline{i}}/\underline{s}_{\underline{i}} \ (\underline{c}_{\underline{i}}, \underline{c}_{\underline{j}}) = 1 + \Sigma_{\underline{j}} \ \underline{k}^{\underline{c}}_{\underline{i}\underline{j}} \underline{c}_{\underline{j}} \qquad (18)$$

EXPERIMENTAL

Different types of experiment were performed to determine $\underline{k}_{12}$ and $\underline{k}_{21}$ respectively when $\underline{s}^0_1 > \underline{s}^0_2$. If a ternary solution is subjected to sedimentation in a standard cell, the boundary of the fast component moves in the ternary solution of known composition. Thus from experiments in the standard cell the value of k_{12} is obtained.

In order to determine k_{21}, the boundary of the slow component must be made to move in the ternary solution. For that purpose a two-layer cell was employed with a ternary solution on the bottom and a binary solution of the fast component on top.

In both types of experiment the drop in sedimentation velocity at the boundary between the two layers gives rise to a concentration difference of the common component in the two layers. These effects are called Johnston/Ogston effects (ref. 10).

Johnston/Ogston effects (ref. 18)

In both types of experiment the cell contains a ternary layer, with concentrations c_i^t and c_j^t, on the bottom and a binary layer, with concentration c_i^b, on top of the ternary layer.

The fluxes of component i at both sides of the interlayer boundary must be equal in a stationary state, thus

$$c_i^b (s_i^b - s_j^t) = c_i^t (s_i^t - s_j^t) \quad . \tag{19}$$

In the standard cell the binary layer contains the slow component, so

$$s_j^t > s_i^b > s_i^t \text{ leading to } c_i^b > c_i^t \quad . \tag{20}$$

In the two layer cell the binary layer contains the fast component

$$s_i^b > s_i^t > s_j^t \text{ leading to } c_i^b < c_i^t \quad . \tag{21}$$

The increment of c_i^b in the standard cell can be considered to originate from molecules i which are overtaken by the interlayer boundary. The binary layer with the stationary concentration of c_i^b comes into existence and grows during the sedimentation process.

The decrease in c_i^b as compared with c_i^t in the two-layer cell can be considered to originate from molecules overtaking the interlayer boundary. In this case, with given c_i^t, c_i^b must be chosen such that the number of molecules overtaking the boundary exactly compensates the number of molecules withdrawing from the boundary in the ternary solution. In case c_i^b and c_i^t do not match, either a third layer will occur (if c_i^b is too small) or an instability with convection if c_i^b is too large, while in both cases the measured values of s_i^t will be in error.

RESULTS

Fig.3 summarise the values of k_{ij} measured as indicated. The dashed straight lines in Fig.3 represent theoretical values of k_{ii} of Pyun and Fixman (ref. 11) and Freed (ref. 12) and Edwards respectively, the latter predicting equality of k_{ii} and $[\eta]_i$.

The solid line indicated by $i=j$ records the experimental values of k_{ii}. The other curves in Fig.3 show the values of k_{ij} as functions of s_j^0. It is seen that k_{ij} increases with increase in M_j with a single exception (M_i=110,000 M_j = 2.2 x 10^6), and with increase in M_i with two exceptions (M_j=110,000, M_i=700,000 and M_j=411,000, M_i=1.26 x 10^6).

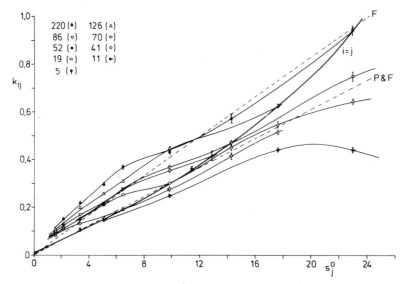

Fig. 3. Interaction constants k_{ij}^c as functions of s_j^0 (refs. 7, 18), values of $10^{-4} M_i$ are indicated. The central curve represents k_{ij}^c. The broken lines are theoretical curves of Pyun and Fixman (ref. 11) and Freed (ref. 12).

The shape of the plots of k_{ij} versus s_j^0 are irregular in general, suggesting the combined effect of several phenomena. A discussion of the results will be given in the following sections.

DISCUSSION

Hard spheres

For the interpretation of hydrodynamic properties of macromolecules the model of the equivalent spheres has been successful. For hard spheres hydrodynamic properties can be calculated as functions of their radius or volume. The equivalent sphere of a macromolecule is a sphere of such a size that it has the same hydrodynamic property as the macromolecule. Thus from the friction coefficient f_0 of a macromolecule in very dilute solution a friction equivalent radius R_f can be calculated by using Stokes' equation

$$f_0 = 6\pi\eta R_f \quad .$$ (22)

Similarly a viscosity equivalent volume $V_\eta = \dfrac{4\pi}{3} R_\eta^3$ can be calculated from the viscosity index $[\eta]$, using Einsteins' equation $[\eta]c = 2.5 \phi$ so

$$V_\eta = \frac{[\eta] M}{2.5 N_{Av}}$$ (23)

Also from the friction interaction coefficient, k_{ii}, a characteristic volume V_k can be calculated as follows. For hard spheres Burgers (ref. 13) calculated

$$k^\phi = \lambda_1 + \lambda_2 = 6.88 \quad .$$ (24)

Here $\lambda_1 = 5$ represents the contribution of backflow and 1.88 is the contribution of reflections of flow perturbations. Later more exact calculations of Batchelor (ref. 15) yield $k^\phi = 6.55$, while Peeters and Staverman (ref. 14) from different refinements of Burgers' calculations obtained values 6.8, 6.42 and 6.53. We will keep to Batchelors' value

$$k^\phi = 6.55 \quad .$$ (25)

In a similar way to the relation between V_η and $[\eta]$ we derive

$$V_k = \frac{k^c M}{6.55 \, N_{Av}} = \frac{4\pi}{3} R_k^3 \quad .$$ (26)

The interaction constant $\underline{k}_{ij}^{\phi}$ between hard spheres of different radii with ratio $\epsilon = \underline{R}_i/\underline{R}_j$ and different velocities with ratio $\underline{v}_j^0/\underline{v}_i^0$ has also been calculated by Peeters (refs. 7, 14) by means of the reflection method. His final equation is complicated. It can be written

$$\underline{k}_{ij}^{\phi} = \frac{\underline{v}_j^0}{\underline{v}_i} \underline{F}_1(\epsilon) + \underline{F}_2(\epsilon) \tag{27}$$

with

$$\epsilon = \underline{R}_i/\underline{R}_j \quad . \tag{28}$$

$\underline{F}_1(\epsilon)$ and $\underline{F}_2(\epsilon)$ are sums of functions of ϵ, values of which have been tabulated by Peeters. $\underline{F}_1(\epsilon)$ and $\underline{F}_2(\epsilon)$ are monotonically increasing functions of ϵ. A crude approximation of $\underline{F}_1(\epsilon)$ is:

$$\underline{F}_1(\epsilon) = 1.064 \ \epsilon^2 + 2.745 \ \epsilon + 1.047 \tag{29}$$

with $\underline{F}_1(1) = 4.844$, and $\underline{F}_2(\epsilon)$ slowly increasing from $\underline{F}_2(0.1) = 0.43$, $\underline{F}_2(1) = 1.958$ to $\underline{F}_2(10) = 2.32$, so that

$$\underline{k}_{ii}^{\phi} = \underline{F}_1(1) + \underline{F}_2(1) = 6.802 \tag{30}$$

not far from Burgers' value.

From eqn (27) it is seen that two effects contribute to the interaction between hard spheres; the first term is proportional to $\underline{v}_j^0/\underline{v}_i^0$ and represents the effect of backflow, the second term describes the field distortion. In the limit $\epsilon^{-1} \to 0$ ($\underline{R}_j \to 0$) and $\underline{v}_j^0 \to 0$ Peeters finds

$$\lim \underline{k}_{ii}^{\phi} = 2.5 = \underline{F}_2(\infty) \quad \text{Einstein's value} \quad . \tag{31}$$

The equivalent radii R_f, R_η and R_k

From the measured values of $[\eta]$ and $\underline{s}^0$, given by eqn (14) and (15), and from eqn (9), (10), (22) and (23) the equivalent radii are calculated to be (ref. 16)

$$\underline{R}_\eta = 0.236 \times 10^{-8} \ \underline{M}^{\frac{1}{2}} \tag{32}$$

and

$$\underline{R}_f = 0.219 \times 10^{-8} \ \underline{M}^{\frac{1}{2}} \quad . \tag{33}$$

The fact that $\underline{R}_\eta$ and $\underline{R}_f$ are not very different does not mean that a sphere is a realistic model of most conformations of the macromolecule, nor that most segments are located within distance $\underline{R}_f$ of the centre of gravity. In fact, the radius of gyration, $\underline{R}_g$, of polystyrene molecules at $\underline{T} = \theta$ appears to be (ref. 7)

$$\underline{R}_g = 0.36 \times 10^{-8} \ \underline{M}^{\frac{1}{2}} \tag{34}$$

so many segments are on the average located outside the sphere with $\underline{R} = \underline{R}_f$. The observation that $\underline{R}_\eta$, $\underline{R}_f$ and $\underline{R}_g$ are all strictly proportional to $\underline{M}^{\frac{1}{2}}$ indicates that homologous macromolecules at $\underline{T} = \theta$ behave in friction and in shear flow as congruent bodies. It follows from eqns (15 and 16), if the third term of the l.h.s. of eqn (16) is neglected, that

$$\underline{k}^c = 0.0515 \ \underline{M}^{\frac{1}{2}}(1 + 3.3 \times 10^{-4} \ \underline{M}^{\frac{1}{2}}) = 0.0515 \ \underline{M}^{\frac{1}{2}}\underline{Q} \quad . \tag{35}$$

The factor $\underline{Q}$ runs from unity for small $\underline{M}$ to 1.49 for $\underline{M}=2.2 \times 10^6$.
Substituting eqn (35) in eqn (26) we find

$$\underline{R_k} = 0.146 \times 10^{-8} \; \underline{M}^{\frac{1}{2}} \; \underline{Q}^{1/3} \text{ with } 1 < \underline{Q}^{1/3} < 1.14 \qquad . \tag{36}$$

Thus for small $\underline{M}(\underline{Q}=1)$

$$\underline{R_k}/\underline{R_f} = 0.146/0.219 = 2/3 \qquad . \tag{37}$$

While $\underline{R_f}$ is much smaller than the "size" of the molecule, represented by $\underline{R_g}$, $\underline{R_k}$ is again much smaller than $\underline{R_f}$. Not only the model of the equivalent sphere, but also that of the congruent bodies is eliminated by these observations. For equivalent spheres $\underline{R_k}$ should be equal to $\underline{R_f}$, while for equivalent congruent bodies $\underline{R_k}$ should be strictly proportional to $\underline{M}^{\frac{1}{2}}$.

Physical interpretation of the value of k^c

Reduction of backflow and field distortion due to porosity of the macromolecule as compared with the hard sphere cannot explain the small value of $\underline{R_k}$, because this reduction is already taken into account in the value of $\underline{R_f}$. This effect alone should make $\underline{R_k}$ equal to $\underline{R_f}$. Another effect is not accounted for in calculations about hard spheres, but has been envisaged by Pyun and Fixman (ref. 11).

Pyun and Fixman suggest that, particularly in dilute θ-solutions, considerable intertwinement of macromolecules will take place. They consider pairs of intertwined molecules as ellipsoids with dimensions adapted to those of the original equivalent spheres. Since these ellipsoids have twice the mass of a single molecule and a friction much smaller than twice that of a single molecule, a considerable reduction in $\underline{k}^\phi$ as compared with that of the separate spheres can be understood. They derive the equation

$$\underline{k}^\phi = 7.01 - \underline{F} \; (\underline{A_2}) \qquad . \tag{38}$$

Here $\underline{F} \; (\underline{A_2})$ is a function of the virial coefficient, $\underline{A_2}$, and 7.01 is the value (due to an error they actually write 7.16), comparable to the values of Burgers and Batchelor for hard spheres.

For θ-solutions $(\underline{A_2}=0)$ $\underline{F} \; (0) = 4.16$, 4.64 and 4.95 (corrected values of Peeters (refs. 7, 19)), depending on the shape of the ellipsoids. This corresponds to ratio´s $\underline{k}/\underline{s}^0 = 3.53$, 2.80 and 2.36, in reasonable agreement with eqn (16). In view of the crude and unrealistic assumptions in the treatment of Pyun and Fixman we are inclined to consider the quantitative agreement as fortuitous. However, the basic idea, that in bad solvents intertwining of macromolecules gives rise to a considerable reduction of friction, appears sound. Elsewhere (ref. 17) we have, retaining the basic idea of Pyun and Fixman, developed a more realistic and simpler model of intertwining macromolecules. The theory of Pyun and Fixman contains two essential elements, expressed in two basic quantities.

The first element is the idea, that the average velocity of an intertwining pair of molecules is, averaged over all possible pairs, larger than that of a single molecule. This can be expressed in a quantity $\underline{r}$ the velocity ratio of double molecules (d) and single molecules (s):

$$\underline{r} = \underline{v}_d^0 / \underline{v}_s^0 \qquad . \tag{39}$$

The second element is the probability of intertwining as a function of concentration and of molecular weight; this can be expressed by a quantity $\underline{x}$, defined by

$$\underline{R}_{int} = \underline{x}\underline{R}_f \qquad . \tag{40a}$$

Here $\underline{R}_{int}$ is the average distance of approach between two macromolecules within which intertwinement occurs. Particularly in -solutions, where the distribution function $\underline{g}(\underline{R}_{ij})$ of the distance $\underline{R}_{ij}$ between two molecules is supposed to be independent of $\underline{R}_{ij}$, the probability of intertwining can be derived immediately from $\underline{R}_{int}$ and, therefore, from $\underline{x}$. $\underline{R}_{int}$, an intermolecular distance, should not be confused with $\underline{R}_k$, defined by eqn (26). $\underline{R}_k$ is not a distance but a friction-interaction-equivalent radius.

In their model Pyun and Fixman, considering the single molecules as real spheres, assume that intertwinement begins when these spheres touch each other. This implies the unjustified assumption

$$\underline{R}_{int} = 2\underline{R}_f \text{ (ref. 11)} \tag{40b}$$

Since the molecules are almost never spherical and the radius $\underline{R_f}$ is only an equivalent radius, there is no reason why intertwinement should begin at $2\underline{R_f}$. On the contrary, the fact that $\underline{R_g}=1.6 \; \underline{R_f}$ justifies the expectation that $\underline{R}_{int}$ will be larger.

In our alternative model (ref. 17) we consider a double molecule as a molecule from the same homologous series with twice the number of segments of a single molecule. This yields immediately a probable value of $\underline{r}$. We have

$$\underline{M}_d = 2\underline{M}_s; \quad \underline{R}_d = 2^{\frac{1}{2}}\underline{R}_s; \quad \underline{f}_d = 2^{\frac{1}{2}}\underline{f}_s \quad \text{and} \quad \underline{s}_d^0/\underline{s}_s^0 = \underline{v}_d^0/\underline{v}_s^0 = 2^{\frac{1}{2}} = \underline{r} \quad . \tag{41}$$

Admittedly a cluster of 2 molecules with $\underline{N}$ segments each is not identical with a single molecule with $2\underline{N}$ segments. But if the average radius of the cluster should be somewhat smaller than $2^{\frac{1}{2}}\underline{R}_s$, the effect of this deviation on velocity will be compensated, partly or wholly, by the effect of the higher internal segment density. This justifies confidence in eqn (41). From the experimental values of $\underline{k}^c$ in eqn (35) together with eqn (40a) and (41) an experimental value of $\underline{x}$ can be derived. For small $\underline{M}$ we find $\underline{x} = 2.23$, for $\underline{M} = 2.2 \times 10^6$, $\underline{x} = 2.06$. Thus $\underline{R}_{int}$ is indeed somewhat larger than $2\underline{R}_f$, but still smaller than $\underline{R}_g$.

The decrease of $\underline{x}$ with increase of $\underline{M}$ is a consequence of the non-linearity of $\underline{k}^c$ versus $\underline{M}^{\frac{1}{2}}$. This variation cannot be explained by the model of Pyun and Fixman, because in that model $\underline{x}$ and $\underline{r}$ are independent of $\underline{M}$. In our alternative model the origin of the variation of $\underline{x}$ is clear; the probability of intertwining will depend on the product of local segment densities of 2 molecules in regions between these molecules. With increase of $\underline{M}$ the volumes of spheres with radii $\underline{R}_g$ or $\underline{R}_f$ increase with $\underline{M}^{3/2}$, whereas the number of segments increases with $\underline{M}$. Consequently the segment densities at equivalent distances decrease with increase of $\underline{M}$. In order to acquire the same product of local densities, molecules of high molecular weight must interpenetrate relatively deeper than small molecules. This is revealed by decrease of $\underline{R}_{int}$ and $\underline{x}$ for large $\underline{M}$.

Peeters and Smits (ref. 19) explain the non-linearity of $\underline{k}$ versus $\underline{M}^{\frac{1}{2}}$ by assuming that $\underline{r}$ depends on $\underline{M}$, as a consequence of non-uniform interpenetration of macromolecules at $\underline{T} = \theta$ (Olaj (ref. 20)). They adhere to the unjustified assumption eqn (40b).

The fact that the value of $\underline{x}$, calculated from the experimental value of $\underline{k}^c$, is not far from 2, and that for the ellipsoids of Pyun and Fixman the value of $\underline{r}$ is not far from $2^{\frac{1}{2}}$, explains the reasonable agreement of their model with experiment.

The values of k_{ij}

Three effects dominate the interaction between like molecules and determine the value of $\underline{k}_{ii}$. These effects contribute also to the values of $\underline{k}_{ij}$. These effects are: backflow and field distortion, porosity and intertwinement. Backflow and field distortion are responsible for the interaction between hard spheres. For spheres of different radii, $\underline{R}_i/\underline{R}_j = \;$, $\underline{k}_{ij}(\;)$ is given by eqn (27) with eqns (29) and (31).

The effect of porosity is also manifest in the friction of a single molecule and is accounted for by assigning the radius $\underline{R} = \underline{R}_f$ to the equivalent spheres. For a theoretical description of the effect of intertwining between molecules of different size, values of x_{ij} and Q_{ij} in eqns (40) and (35) must be known. This requires an extension of the theory which will be the subject of separate papers (ref. 17).

A crude method to take account of the effect of intertwining on the value of $\underline{k}_{ij}$ consists in dividing the experimental values of $\underline{k}_{ij}^c$ by $\underline{k}_{jj}^c$ or by $\underline{k}_{ii}^c$. The characteristic experimental quantities are then

$$\underline{G}_{exp} = \underline{k}_{ij}^c / \underline{k}_{jj}$$

$$\underline{H}_{exp} = \underline{k}_{ij}^c / \underline{k}_{ii}^c \quad . \tag{43}$$

In order to compare experimental and theoretical quantities, experimental values of $\underline{k}_{ij}^c$ must be converted into values of $\underline{k}_{ij}^\phi$ by means of the equation

$$\underline{k}_{ij}^c = \frac{\underline{N}_{Av}\underline{V}_j}{\underline{M}_j} \underline{k}_{ij}^\phi \quad . \tag{44}$$

Here $\underline{V}_{\underline{j}}=(\underline{V}_{\underline{ij}})_k$, the interaction-equivalent volume, comparable to $\underline{V}_k$ in eqn (26). It depends on unknown values of $\underline{x}_{\underline{ij}}$ and $\underline{Q}_{\underline{ij}}$ in the model of intertwining molecules.

If we make the assumption that $\underline{V}_{\underline{j}}$ has the same value in $\underline{k}^c_{\underline{ij}}$ and $\underline{k}^c_{\underline{jj}}$, we arrive at the equation

$$\frac{\underline{k}^c_{\underline{ij}}}{\underline{k}^c_{\underline{jj}}} = \frac{\underline{k}^\phi_{\underline{ij}}}{\underline{k}^\phi_{\underline{jj}}} = \frac{\varepsilon^{-1}\underline{F}_1(\varepsilon) + \underline{F}_2(\varepsilon)}{6.8} \equiv \underline{G}(\varepsilon) \quad . \tag{45}$$

Here we have used the relations $\underline{v}^0_{\underline{j}}/\underline{v}^0_{\underline{i}} = \varepsilon^{-1}$ and $\underline{k}_{\underline{jj}} = \underline{F}_1(1) + \underline{F}_2(1) = 6.8$. Similarly with $\underline{k}^c_{\underline{ii}} = \underline{N}_{Av}\underline{V}_{\underline{i}}\underline{k}^\phi_{\underline{ii}}/\underline{M}_{\underline{i}}$ we find

$$\frac{\underline{k}^c_{\underline{ij}}}{\underline{k}^c_{\underline{ii}}} = \frac{\underline{M}_{\underline{i}}\underline{V}_{\underline{j}}}{\underline{M}_{\underline{j}}\underline{V}_{\underline{i}}} \frac{\underline{k}^\phi_{\underline{ij}}}{\underline{k}^\phi_{\underline{ii}}} = \frac{\varepsilon^{-2}\underline{F}_1(\varepsilon) + \varepsilon^{-1}\underline{F}_2(\varepsilon)}{6.8} \equiv \underline{H}(\varepsilon) \quad . \tag{46}$$

Here we have made the additional assumption $\underline{M}_{\underline{i}}\underline{V}_{\underline{j}}/\underline{M}_{\underline{j}}/\underline{V}_{\underline{i}} = \varepsilon^{-1}$. In Figs. 4 and 5 plots are shown of $\log \underline{H}(\varepsilon)$ and $\log \underline{G}(\varepsilon)$ versus ε together with the experimental values of $\log \underline{H}_{exp}$ and $\log \underline{G}_{exp}$ respectively.

Fig. 4 shows that the general trend and the order of magnitude of the values of $\underline{H}(\varepsilon)$ and $\underline{H}_{exp}$ are not very different.

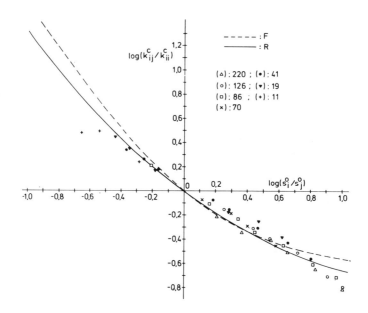

Fig. 4. $\log \underline{H}_{exp} = \log(\underline{k}^c_{\underline{ij}}/\underline{k}^c_{\underline{ii}})$ versus $\log \varepsilon = \log(\underline{s}^0_{\underline{i}}/\underline{s}^0_{\underline{j}})$, (ref.18). The drawn curve is theoretical: $\log \underline{H}(\varepsilon)$, eqn (46). The definitions of $F(\varepsilon)$ and $R(\varepsilon)$ are given in eqns (27) and (28).

This suggests that interaction between macromolecules of different size originates mainly from backflow and field distortion, the only effects operating between hard spheres. However, Fig. 5 contradicts this conclusion. Here we see, that large molecules are more strongly retarded by small molecules than the corresponding spheres (larger than $\underline{k}_{\underline{ij}}$ for $\varepsilon > 1$), while small molecules are less retarded or even accelerated by large molecules (small $\underline{k}_{\underline{ij}}$ for $\varepsilon < 1$). However, the experimental points in the figures show large and irregular deviations from the theoretical curves, indicating that other effects besides those mentioned are operating between molecules of different size and velocity. The figures show conclusively, that $\underline{G}_{exp}$ and $\underline{H}_{exp}$ are not unique functions of ε as they should be if homologous macromolecules at $\underline{T} = \theta$ behaved as congruent bodies.

In a more detailed analysis of the available data the following effects should be taken into account:

(i)   the effect of difference in size upon $(R_{int})_{ij}$ or x_{ij},

(ii) the effect of large M_i and/or M_j on the factor Q in eqn (35),

(iii) the effect of the difference in velocity on the nature and frequency of close contacts ("collisions") between molecules of different size.

Whereas intertwining accelerates both molecules, the slow ones more than the fast ones, collisions without intertwining will possibly retard fast molecules.

Although the measurements comprise no less than 40 pairs and 9 binary solutions, probably the amount of data is insufficient to allow definite conclusions about the contributions of the different effects.

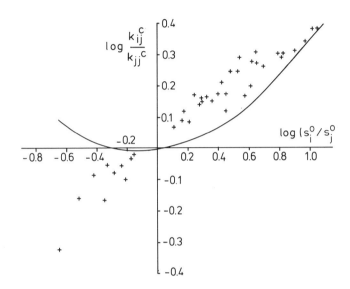

Fig. 5. Log G_{exp} = $\log(k_{ij}^c/k_{jj}^c)$ versus log ϵ = $\log(s_i^0/s_j^0)$. The drawn curve is theoretical: log $\underline{G}$ (ϵ), eqn (45).

CONCLUSIONS

I have tried to make clear that, while determinations of molecular weight and MWD´s can be reliably performed with other methods, the ultracentrifuge remains a very useful, if not unique, source of information about interaction between macromolecules. In sedimentation processes of ternary solutions of macromolecules, four different mechanisms of interaction are operating, each characteristic for the size and structure of the molecules under consideration. These mechanisms are backflow, field distortion, intertwining and collisions. Results obtained so far, are definitely at variance with the model of the equivalent sphere and also with that of the congruent bodies, including the ellipsoids of Pyun and Fixman. I have suggested qualitative explanations for some results of the reported experiments but a quantitative analysis of the various effects will remain a challenge to polymer scientists for quite some time and, I hope, a stimulus to many more experiments with the ultracentrifuge.

REFERENCES

1. The Svedberg and H. Rinde, J. Am. Chem. Soc. **45**, 943 (1923).
2. The Svedberg and J.B. Nichols, J. Am. Chem. Soc. **45**, 2910 (1923).
3. H. Fujita, Foundation of Ultracentrifugal Analysis, Wiley, London (1975).
4. R.L. Baldwin, J. Phys. Chem. **58**, 1081 (1954).
5. J.J.H. Mulderije, Thesis Leiden (1978).
6. J.J.H. Mulderije, European Polymer J. **17**, 807 (1981).
7. F.A.H. Peeters, Thesis Leiden (1979).
8. F.A.H. Peeters and A.J. Staverman, Proc. Kon. Ned. Akad. Wet. B **84**, 161 (1981).

9. N. Ho-Duc, H. Daoust and F. Bisonnette, Can. J. Chem **50** 305 (1972).
10. J.P. Johnston and A.G. Ogston, Trans. Faraday Soc. **42**, 789 (1946).
11. C.W. Pyun and M. Fixman, J. Chem. Phys. **41** 937 (1964).
12. K.F. Freed, J. Chem. Phys. **65**, 4103 (1976).
13. J.M. Burgers, Proc. Kon. Akad. Wet. A´dam, **44**, 1045, 1177, (1941); **45**, 126 (1942).
14. F.A.H. Peeters and A.J. Staverman, Proc. Kon. Ned. Akad. Wet. B **83**, 209 (1980).
15. G.K. Batchelor, J. Fluid Mech. **52**, 245 (1972).
16. J.J.H. Mulderije, Macromolecules. **13**, 1207 (1980).
17. A.J. Staverman, European Polymer Journal **21**, 175 (1985); F.A.H. Peeters, Bull. Soc. Chim. Belg. To be published.
18. F.A.H. Peeters and A.J. Staverman, Proc. Kon. Ned. Akad. Wet. B **85**, 273 (1982), **86**, 89 (1983).
19. F.A.H. Peeters and H.J.E. Smits, Bull. Soc. Chim. Belg. **90**, 111 (1981).
20. O.F. Olaj and K.H. Pelinka, Makromol. Chem. **177** 3413 (1976).

15. Sedimentation Analysis of Proteins

Hidematsu Suzuki

Institute for Chemical Research
Kyoto University, Uji, Kyoto-Fu 611, Japan

Abstract – Early developments in the use of ultracentrifuge and in
ultracentrifugation remind us that Svedberg and his colleagues must have
thoroughly understood the potentials of the ultracentrifuge and that
they established almost all basic methods in ultracentrifugation, with
the exception of the density gradient technique. Their theories were
often restricted to ideal, homogeneous systems. This was quite natural,
because it was not yet known in those days how to express the activity
coefficient in terms of molecular parameters. In two topics selected,
molecular weight determination and sedimentation analysis of reacting
systems, recent advances in treatments of the effects of nonideality and
heterogeneity are described. In the former, the importance of low-speed
operations is illustrated especially for samples having broad distribution. Extension of the upper limit of molecular weight determinable
with the ultracentrifuge is also mentioned. In the latter, there are
two approaches: one involves the average molecular weight for graphical
analysis, while the other is a direct computer analysis of the concentration vs. radial position curve observed. An example of analysis
for each approach is presented to show the recent state of the study.

INTRODUCTION

In the large field indicated by the above title, only two topics will be covered in this
account. They are molecular weight determination and sedimentation analysis of reacting
systems. Before discussing these topics, it seems worthwhile to recall the early developments in the use of the ultracentrifuge and in ultracentrifugation. Table 1 lists ten
selected developments (refs. 1 - 11), which are related to the discussion below. This is
not an exhaustive list for the period covered. Some other significant contributions are
deleted, simply because they are concerned with transport phenomena or refinements of the
instruments.

It should be noted first from the table that Svedberg did construct both low-speed and
high-speed ultracentrifuges in the early days of his study of the ultracentrifuge (refs. 1,
7). The terms low-speed and high-speed ultracentrifuges are little used today, but they are
necessary to distinguish the two instruments. Both are equally important like each wheel of
a bicycle. These instruments which give centrifugal fields varying from 60 to 300,000 times
gravity became applicable to different research problems. Without such versatility of his
instruments, it would have been impossible to determine the molecular weights of very large
proteins like Helix pomatia haemocyanin at a speed of 1,180 rpm (ref. 9) and to establish
the sedimentation equilibrium state of aqueous solutions of inorganic salts like LiCl at a
speed of 45,000 to 60,000 rpm (ref. 10). Accordingly, it is a mistake to say that the first
crude, low-speed machines were evolved from the massive, high-speed instruments. When one
looks back upon the very slow development of the low-speed type of ultracentrifuge, the fact
that two different instruments were built in Uppsala cannot be stressed enough.

Secondly Svedberg and his colleagues developed almost all the ultracentrifugal methods
currently used, with the exception of the density gradient technique in its many variations
(refs. 12, 13). Svedberg and Rinde measured the particle size in gold sols from sedimentation velocity runs and discussed its heterogeneity (ref. 1). Svedberg and Fåhraeus first
determined the molecular weight of haemoglobin by sedimentation equilibrium runs, finding
that this protein was homogeneous and consisted of four subunits (ref. 3). Successively,
Svedberg and Nichols confirmed the homogeneity of haemoglobin from sedimentation velocity
runs and calculated its molecular weight from sedimentation and diffusion coefficients by
means of the so-called Svedberg equation (ref. 8). Both sedimentation velocity and
sedimentation equilibrium are still two principal methods in modern ultracentrifugation.

Tiselius discussed the (primary) charge effect and the association - dissociation
reaction of proteins in sedimentation equilibrium (ref. 6). From this work, the current
extensive ultracentrifuge studies on the self-association of proteins originated. However,
there was no followup until 1961 when Squire and Li carried out sedimentation equilibrium

Table 1. Early developments in the ultracentrifuge and in ultracentrifugation

Developments	Researchers	Ref.

Low-speed ultracentrifuge with gear drive,
    definition of ultracentrifuge,
    sedimentation of gold sols,
    square dilution law for a sector cell
	Svedberg & Rinde
	(1924)
		(1)

Thermodynamic theory of molecular weight
    measurements by sedimentation
    equilibrium
	Svedberg (1926)	(2)

First measurements of molecular weights
    of proteins (haemoglobin and ovalbumin)
    by sedimentation equilibrium
	Svedberg & Fåhraeus
	(1926); Svedberg &
	Nichols (1926)
		(3)
		
		(4)

Definition of time to attain sedimentation equilibrium under certain conditions	Weaver (1926)	(5)

Thermodynamic theory of sedimentation
    equilibrium, charge effects and
    associating systems
	Tiselius (1926)	(6)

High-speed ultracentrifuge with oil-turbine drive	Svedberg & Lysholm (1927)	(7)

Sedimentation velocity of haemoglobin
    by high-speed ultracentrifuge
    molecular weight from s and D*
	Svedberg & Nichols
	(1927)
		(8)

First sedimentation equilibrium and
    velocity studies on very large
    protein, Helix pomatia haemocyanin
	Svedberg & Chirnoaga
	(1928)
		(9)

Thermodynamics of sedimentation
    equilibrium, activity measurements
    of inorganic salt solutions
	Pedersen (1934)	(10)

Various average molecular weight definitions and calculations	Lansing & Kraemer (1935)	(11)

* s and D: sedimentation and diffusion coefficients, respectively

experiments to elucidate the state of aggregation of ovine pituitary adrenocorticotropin (ACTH) in acid and base solutions (ref. 14). Another important application, due to Pedersen (ref. 10), of sedimentation equilibrium to activity measurements was not extended to heterogeneous systems until 1970 when Scholte determined the thermodynamic parameters of polystyrene solutions over a wide range of concentration (ref. 15).

Lansing and Kraemer demonstrated the feasibility of measuring the number-, weight- and z-average molecular weights of a given heterogeneous sample by sedimentation equilibrium (ref. 11). This is a great advantage of the ultracentrifuge because the usual methods of determining the molecular weight yield only one average value. In addition, since the solute concentration at sedimentation equilibrium changes throughout the solution column, a single ultracentrifuge experiment allows a series of each average molecular weight to be determined over a certain range of the total concentration. This is another advantage of the ultracentrifuge over other instruments, especially in the study of self-association of proteins.

As can be seen from the comments above, Svedberg was bright and so foresighted that he and his colleagues established almost all basic methods in ultracentrifugation. Their theories were often restricted to thermodynamically ideal, homogeneous systems. This is understandable, because many developments in statistical mechanics and solution thermo-dynamics had not yet been accomplished in those days (refs. 16 - 18). Effects of the nonideality of solution and the heterogeneity of solutes were formidable problems, having only recently been to a large extent solved (refs. 19 - 21). Hence, advances in the treat-ments of these effects will be described in the following sections. Our understanding of the sedimentation velocity behaviour in nonideal, heterogeneous systems is still limited, and we will mention it only briefly when necessary.

MOLECULAR WEIGHT DETERMINATION

In protein chemistry, much has changed in the last two decades. Sodium dodecyl sulphate gel electrophoresis is frequently used to estimate the number and rough size of the peptide chains. Even the exact molecular weights of these chains can be chemically determined from the amino acid analysis. Therefore, ultracentrifugation has become a classical method for determining the molecular weights. However, for other macromolecules like poly-saccarides and synthetic polymers, ultracentrifugal methods do not lose their value.

Among those methods, sedimentation equilibrium is most suitable for this purpose. Except the earliest one (ref. 22), the theories due to Svedberg and his colleagues (refs. 2, 6, 10) were based on thermodynamics. They had recognized, as Goldberg did some twenty years later, that at sedimentation equilibrium, the total potential of a component is constant throughout the system. Unfortunately, it was impossible to express the activity coefficient in terms of molecular parameters (pair correlation functions) until the work of McMillan and Mayer (ref. 16) and others (refs. 17, 18). Using these advanced theories, Goldberg succeeded in giving the basis for analytical theory of sedimentation equilibrium (ref. 23).

Fujita (ref. 24) put his theory into a form convenient for practical use, finding that the apparent weight-average molecular weight $\underline{M}_w^{app}$ defined below depends on the initial concentration $\underline{c}_o$ and a generalized speed parameter λ :

$$1/\underline{M}_w^{app} = 1/\underline{M}_w + B_{LS} [1 + \tfrac{1}{12}(\lambda \underline{M}_z)^2 + \ldots] \underline{c}_o + \ldots \tag{1}$$

$$\underline{M}_w^{app} = [\underline{c}(\underline{b}) - \underline{c}(\underline{a})]/\lambda \underline{c}_o \tag{2}$$

$$\lambda = (1 - \bar{v}\rho_o)(\underline{b}^2 - \underline{a}^2)\omega^2/2\underline{RT} . \tag{3}$$

Here, $\underline{M}_w$ and $\underline{M}_z$ are the weight- and z-average molecular weights, respectively; B_{LS} the light-scattering second virial coefficient; $\underline{c}$ the solute concentration; $\underline{a}$ and $\underline{b}$ the radial positions at the top and bottom of a solution column, respectively; $\bar{v}$ the partial specific volume of solutes; ρ_o the solvent density; ω the angular velocity of the rotor; $\underline{RT}$ has the usual significance. Eqn (1) suggests that the double extrapolations should be made first with respect to λ^2 and then with respect to $\underline{c}_o$. For the former extrapolation, runs should be performed at different values of λ. This is easily carried out by changing the rotor speed. However, such experiments are still cumbersome.

Williams et al. (ref. 25) derived, from Goldberg´s theory, the following important relation

$$1/\underline{M}^{app} = 1/\underline{M} + \underline{B}[\underline{c}(\underline{a}) + \underline{c}(\underline{b})]/2 \tag{4}$$

for two-component systems of homogeneous solutes and solvent provided that the second virial coefficient $\underline{B}$ is sufficient to describe the nonideality. The new concentration variable in this equation is usually called the average concentration $\bar{c}$ and the apparent molecular weight can be calculated by eqn (2). When solutions of different concentrations are examined, only single rotor-speed results are required for this analysis procedure. As is obvious from eqn (4), no λ-effects appear in plots of $1/\underline{M}^{app}$ against $\bar{c}$. Extensive application of eqn (4) to heterogeneous systems can be seen in the literature, with a less sound theoretical basis (refs. 26, 27).

Indeed, from the late 1960´s to the early 1970´s, there have been published many papers on the establishment of the standard method of sedimentation equilibrium analysis in non-ideal, heterogeneous systems (refs. 26 - 31). Their methods are essentially variations of either eqn (1) or eqn (4). They are individually successful but of limited applicability. Neither eqn (1) nor eqn (4) are found adequate for highly heterogeneous samples.

The second term on the r.h.s. of eqn (1) is a form obtained after the expansion of exponential and the truncation of the series yielded. The original form of the factor including λ reads

$$\frac{\lambda \underline{M}_i}{\exp(\lambda \underline{M}_i) - 1} \quad \frac{\lambda \underline{M}_k}{\exp(\lambda \underline{M}_k) - 1} \quad \frac{\exp[\lambda (\underline{M}_i + \underline{M}_k)] - 1}{\lambda(\underline{M}_i + \underline{M}_k)} \tag{5}$$

$$= \begin{cases} 1 + (1/12)\lambda^2 \underline{M}_i \underline{M}_k + \ldots & (\lambda \underline{M}_i \ll 1, \ \underline{i} = 1, 2, \ldots, \underline{q}) & (5a) \\ \lambda \underline{M}_i \underline{M}_k/(\underline{M}_i + \underline{M}_k) & (\lambda \underline{M}_i \gg 1) & (5b) \end{cases}$$

If all the $\underline{q}$-components satisfy the condition $\lambda \underline{M}_i \ll 1$, then the expansion into series may be allowed, yielding the same results as Fujita´s. It should be noted here that the condition $\lambda \underline{M}_i$ is more restrictive than the condition often experimentally set, $\lambda \underline{M}_w \leq 1$. Even though the latter condition is fulfilled experimentally, obviously higher molecular-weight components cannot satisfy the former one. On the other hand, for extremely heterogeneous samples like polycondensates (refs. 19, 20), eqn (5b) will hold practically regardless of the rotor speed. This asymptotic relation is useful because the single extrapolation of $1/M_w^{app}$ with respect to λ directly determines the M_w value.

The aforementioned features of eqn (5) have already been demonstrated with some experimental results and numerical calculations (ref. 19). Detailed discussion can be found there. Here, it is repeated that for nonideal, heterogeneous systems, experiments should be carried out at lower and different rotor speeds. An example is given in Fig. 1 to emphasize the usefulness of low rotor speeds.

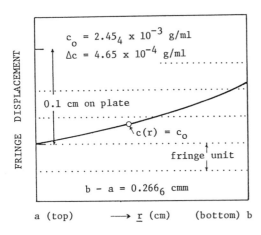

Fig. 1. Equilibrium concentration distribution of a binary polystyrene mixture in methyl ethyl ketone at $30^{\circ}C$ and at 3,189 rpm. For details, see the text.

The sample used is a binary mixture of standard polystyrene having narrow distribution (Pressure Chemical, Pittsburgh, PA): the molecular weight and weight fraction are, respectively, 1.0×10^4 and 0.809_6 for one component and 41.1×10^4 and 0.190_4 for the other. Indeed, this rotor speed of 3,189 rpm is so low that λM_w takes a value of 0.20 and the concentration ratio $c(b)/c(a)$ is found to be only 1.20. However, the fringe shift itself is, as can be seen from its shape, readable with precision. Accordingly, the calculated apparent molecular weight is reliable and close to the true value at a given concentration.

Customarily, it is recommended that the rotor speed be adjusted so that at equilibrium the $c(b)/c(a)$ ratio is somewhere between 2 and 5. This is quite correct for samples having narrow distribution. Pedersen (ref. 32), proposing such an empirical rule for the rotor speed, remarked

> "This, however, is to be regarded as an average empirical rule. If we could choose all the variables (concentration, cell thickness, scale distance, light absorption (or specific refractive increment), speed of the centrifuge, etc.) arbitrarily, a low speed would be preferable, as under this condition the concentration distribution in the cell could be most accurately determined."

Unfortunately, this proviso did not seem to be fully appreciated until the work of Gordon et al. (ref. 19).

Fujita's equation for M_w, eqn (1) has been generalized recently, the formal way of determining the higher average molecular weights like M_z and M_{z+1} being presented (ref. 21). In addition, the number-average molecular weight M_n of a high molecular weight ($M_w = 2.7 \times 10^7$) polystyrene sample was determined as 20 million by meniscus sedimentation equilibrium (ref. 33) with a new extrapolation method (ref. 34).

Apart from the theoretical developments described above, there exists another theory in which the rotor-speed dependence of the apparent molecular weight is taken into account. In each theory of this type, the concentration variable modified only by the experimental quantities is utilized, instead of c_o and λ, to eliminate both the concentration and rotor-speed dependences by a single extrapolation. One was proposed by Van Holde and Williams (ref. 35) in 1953 and another is due to Scholte (ref. 36), who derived equations both for M_w and M_z. As Scholte's work was not published, the equation for M_w is reproduced below.

$$1/M_w^{app} = 1/M_w + (B + \bar{v}/M_w)c_o \int_o^\epsilon (dc/dx)^2 dx / [\int_o^\epsilon (dc/dx)\, dx\,]^2 \qquad (6)$$

Table 2. Molecular weights of some bacteriophages determined by
ultracentrifugal methods

Specimen	Method	$\underline{M} \times 10^{-6}$		Ref.
T7 phage	Meniscus depletion	49.4	(25.3)	(37)
T7 DNA	CsCl Density gradient	24.8		(38)
T5 DNA	CsCl Density gradient	68.7		(38)
T4 DNA	CsCl Density gradient	113		(38)
T7 phage	The Svedberg eqn	50.4	(25.8)	(39)
T5 phage	The Svedberg eqn	109.2	(67.3)	(39)
T4 phage	The Svedberg eqn	192.5	(105.7)	(39)

Values in parentheses are molecular weights for these phage DNA's
calculated from their content.

$$\underline{x} = (\underline{r}^2 - \underline{a}^2)/(\underline{b}^2 - \underline{a}^2) \tag{7}$$

where $\underline{r}$ is the radial distance to a given point in the solution column.

Owing to such actual limitations as the unchangeable buoyancy factor and the lowest rotor-speed available without any precession, there is an upper limit to the molecular weight determinable by sedimentation equilibrium. Roughly speaking, this is 2 – 5 million in the conventional sedimentation equilibrium method. If a given specimen is surely homogeneous in molecular weight, then the meniscus depletion method (ref. 33) can be used without difficulty for analysis. With this method, the upper limit has been extended to about 50 million (ref. 37). Another possibility is to apply the CsCl density gradient sedimentation equilibrium method (ref. 13). It has been applied to some bacteriophage DNA's (ref. 38) but analysis of the results is rather involved. Recently, the diffusion coefficient can be measured quickly and accurately in dilute solutions by means of quasi-elastic light scattering (ref. 39). Accordingly, the classical Svedberg equation below

$$\underline{M} = \underline{RT}\underline{s}^o/\underline{D}^o(1 - \bar{\underline{v}}\rho_o) \tag{8}$$

is frequently and successfully applied to homogeneous particles (ref. 40). Here, $\underline{s}^o$ and $\underline{D}^o$ are the sedimentation and diffusion coefficients at infinite dilution, respectively. Table 2 lists some examples of higher molecular weights determined by the aforementioned ultracentrifugal methods. Since the sedimentation coefficient of such high molecular-weight bacteriophages shows the rotor-speed dependence (ref. 41), the experimental condition proposed by Zimm (ref. 42) should be kept to obtain the right value of $\underline{s}^o$.

SEDIMENTATION ANALYSIS OF REACTING SYSTEMS

The concentration gradient of a nonreacting homogeneous solute in an ideal two-component system is given by

$$d \ln \underline{c}(\underline{r})/d\underline{r}^2 = \underline{M}(1 - \bar{\underline{v}}\rho_o)\omega^2/2\underline{RT} \quad . \tag{9}$$

If each solute species carries the $\underline{z}$ sites ionizable in a given solvent, then the r.h.s. of eqn (9) must be modified by a factor, $(\underline{z} + 1)^{-1}$. This means that the molecular weight calculated from the concentration gradient is reduced to one $(\underline{z} + 1)$th of the true value. This is the primary charge effect first recognized by Tiselius (ref. 6). The effect can be repressed, as was theoretically disclosed by Lamm (ref. 43), by adding an excess of a supporting electrolyte. The residual (secondary) charge effect is to be eliminated by dialysis (ref. 44). Hence, the true molecular weight of a synthetic or natural polyelectrolyte can be obtained unambiguously from sedimentation equilibrium measurements.
   For an ideal, but heterogeneous system, the average molecular weights can be estimated, at any point in the solution column, by use of the relations below.

$$\underline{M}_n = \underline{c}(\underline{r})/[\int d\underline{c}(\underline{r})/\underline{M}_w(\underline{r})] \tag{10a}$$

$$\underline{M}_w = [2\underline{RT}/(1 - \bar{\underline{v}}\rho_o)\omega^2] [d \ln \underline{c}(\underline{r})/d\underline{r}^2] \tag{10b}$$

$$\underline{M}_z = [2\underline{RT}/(1 - \bar{\underline{v}}\rho_o)\omega^2] [d \ln \underline{c}\underline{M}_w(\underline{r})/d\underline{r}^2] \tag{10c}$$

To perform the integration appearing on the r.h.s. of eqn (10a), the concentration must be zero at a certain point, usually, at the top of a solution column. The meniscus depletion sedimentation equilibrium allows such a condition to be met experimentally, and the method has frequently been used for the study on self-association proteins.

Indeed, solutions of most globular proteins are close to ideal, but in general, protein solutions are more or less nonideal due to excluded volume and charge repulsion. In such cases, the molecular weights calculated from eqns (10a – c) are merely the apparent ones. The contribution of the second virial coefficient $\underline{B}$ in the relations below must be accounted for. For a nonideal, homogeneous system with no association, we obtain

$$1/\underline{M}_n^{\,app}(\underline{c}) \;=\; 1/\underline{M} \;+\; (\underline{B}/2)\underline{c}(\underline{r}) \;+\; \cdots \tag{11a}$$

$$1/\underline{M}_w^{\,app}(\underline{c}) \;=\; 1/\underline{M} \;+\; \underline{B}\underline{c}(\underline{r}) \;+\; \cdots \tag{11b}$$

$$1/\underline{M}_z^{\,app}(\underline{c}) \;=\; 1/\underline{M} \;+\; 2\underline{B}\underline{c}(\underline{r}) \;+\; \cdots \quad . \tag{11c}$$

A simple example is shown in Fig. 2, which illustrates the nonideal behaviour of ovalbumin in 0.1 M acetic acid (ref. 45). It can be seen from this figure that two average molecular weights approach the same intercept and their slopes are approximately in agreement with eqns (11a & b). The third plot of $2/\underline{M}_n^{\,app} - 1/\underline{M}_w^{\,app}$ against concentration is based on the relation

$$2/\underline{M}_n^{\,app}(\underline{c}) \;-\; 1/\underline{M}_w^{\,app}(\underline{c}) \;=\; 1/\underline{M} \tag{12}$$

which is derived from eqns (11a & b). This is the basic way of eliminating the nonideality effect in graphical analysis of reacting systems, if the effect can be adequately described by a single virial coefficient.

A reversibly reacting system is distinguishable from a heterogeneous, nonreacting system with ease by sedimentation equilibrium. In a latter system at a fixed temperature, the average molecular weight is a function of the initial concentration before centrifugation. In a former system, on the other hand, the total equilibrium concentration at a given point in the solution column determines the composition, and hence any average molecular weight at that point. A good example of a reacting system is shown in Fig. 3, which illustrates association of chymotrypsinogen A in veronal buffer (ref. 46). This figure shows that data at different initial concentrations are superimposed when plotted against concentration and the molecular weight increases with increasing concentration. The concentration of each species in the aggregates and monomer may be characterized by the equilibrium constant $\underline{K}$. Therefore, determination of the quantities $\underline{M}_1$ (the monomer molecular weight), $\underline{B}$ and $\underline{K}$ are, from data like these, the goal of the analysis of reacting systems.

The methods of analysing a reacting system can be roughly classified into two categories (refs. 47 – 49). One involves the average molecular weights for graphical analysis, while the other is a direct computer analysis of the $\underline{c}(\underline{r})$ versus $\underline{r}$ curve observed.

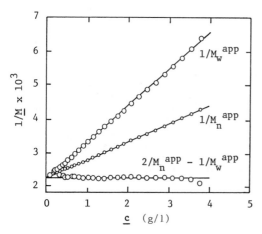

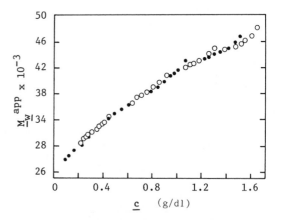

Fig. 2. The nonideal behaviour of oval bumin in 0.1 M acetic acid. Equilibrium ultracentrifugation at 33,450 rpm at 19°C. From Roark and Yphantis (ref. 45).

Fig. 3. Apparent weight-average molecular weight of chymotripsinogen A in veronal buffer (pH 7.9, $\underline{I}$ = 0.03) at 25°C. Different marks are for different initial concentrations. From Hancock and Williams (ref. 46).

(1) Graphical analysis: Although some are valid only for ideal systems, a number of methods have been proposed, which can be roughly divided into three classes. One is the method (refs. 50 - 52) which uses the Steiner equation (ref. 53).

$$\ln \underline{c}_1/\underline{c} = \int_0^{\underline{c}} [(\underline{M}_1/\underline{M}_w) - 1]d \ln \underline{c} \quad . \tag{13}$$

Here, $\underline{c}_1$ is the monomer concentration. Another is based on theoretical dependence of the weight-average molecular weight on concentration (refs. 54, 55). The other makes use of general interelationships between average molecular weights, being represented by the so-called "two-species plot" (refs. 45, 56 - 59). Here, an elegant analysis due to Chun et al. (ref. 59) in terms of a method of the third class is illustrated in Figs. 4 and 5.

Fig. 4 shows plots of $[= (\underline{M}_1/\underline{M}_w{}^{app}) - \ln (\underline{c}_1/\underline{c})^{app}]$ against $[= (2\underline{M}_1/\underline{M}_n{}^{app}) - (\underline{M}_1/\underline{M}_w{}^{app})]$ for glutamic dehydrogenase (ref. 60). In this figure, three solid lineas stand for their theoretical predictions for the monomer-dimer (1 - 2), the monomer-tetramer (1 - 4) and the isodesmic association. The isodesmic association means an indefinite association where the individual equilibrium constants are identical, i.e.,

$$\underline{K}_{12} = \underline{K}_{23} = \cdots = \underline{K}_{i,i-1} = \cdots = \underline{K} \tag{14}$$

in the reaction

$$2 \underline{P}_1 \overset{\rightarrow}{\leftarrow} \underline{P}_2 \qquad\qquad \underline{K}_{12} = \underline{c}_2/\underline{c}_1{}^2$$

$$\underline{P}_2 + \underline{P}_1 \overset{\rightarrow}{\leftarrow} \underline{P}_3 \qquad\qquad \underline{K}_{23} = \underline{c}_3/\underline{c}_2\underline{c}_1$$

$$\cdots \qquad\qquad\qquad \cdots \qquad . \tag{15}$$

$$\underline{P}_{i-1} + \underline{P}_1 \overset{\rightarrow}{\leftarrow} \underline{P}_i \qquad\qquad \underline{K}_{i,\,i-1} = \underline{c}_i/\underline{c}_{i-1}\underline{c}_1$$

It can be seen from Fig. 4 that the experimental points fall almost exactly on the curve for an isodesmic association. Thus the stoichiometry of the present reaction is judged from this type of plot. The next two plots depicted in Figs. 5 (a) and (b) are for the determinations of $\underline{K}$ and $\underline{BM}_1$. These plots resulted in straight lines. From their slopes, $\underline{K}$ and $\underline{BM}_1$ are estimated as 1666 and 6.16, respectively. The linearity of these plots also confirms the fact that no other mode of association than the isodesmic one is operating in the system.

(2) Analysis of the $\underline{c}(\underline{r})$ versus $\underline{r}$ curve: This method seems to have the greater potential because it employs the total information experimentally obtainable. Since Reinhardt and Squire (ref. 61), many procedures for this type of analysis have been examined

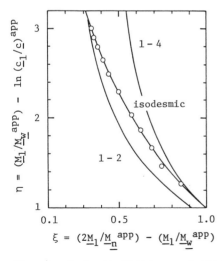

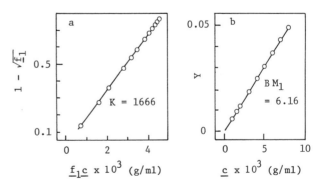

Fig. 4. Analysis of the association of glutamic dehydrogenase due to Chun et al. (ref. 59). The solid lines are functions defined by them. For details, see the text.

Fig. 5. Graphs showing the determinations of (a) $\underline{K}$ and (b) $\underline{BM}_1$ for the data shown in Fig. 4. The quantity $\underline{Y}$ is defined as $(\underline{M}_1/\underline{M}_w{}^{app}) - \sqrt{\underline{f}_1}/(2 - \sqrt{\underline{f}_1})$, where $\underline{f}_1$ is the monomer weight fraction. From Chun et al. (ref.59).

in the literature but none were free of the difficulty in obtaining reliable and unique
results. Recently, Yphantis et al. developed a nonlinear least-square fit of the concentration distribution curve (ref. 62). This method directly determines M_1, B and K from the
interference fringes observed. The numerical procedure and tests of the method with
simulated data can be seen in the original paper. Here we refer to its application to
Limulus haemocyanin (ref. 63).

Purified Limulus haemocyanin exhibits a sedimentation coefficient of approximately 36 S
at pH 6.1 in 0.05 ionic strength phosphate with 0.45 M KCl added.* Since this sedimentation
coefficient increases with haemocyanin concentration, the presence of a reversible reaction
can be expected. Meniscus depletion sedimentation equilibrium experiments were performed on
these solutions with initial concentrations from 0.2 to 2.0 g/l at 26°C and at 5,200 rpm.
The experimental results were processed by NOLIN, the computer program for the nonlinear
least-square fit. The results are listed in Table 3. Unfortunately, three M_w^{app} curves
derived from three different solution columns diverge significantly. This means that under
these conditions, the Limulus haemocyanin system must be heterogeneous. Hence, interpretation of these results are still incomplete. Similar analyses of Limulus haemocyanin under
different conditions were also reported (ref. 63).

In the above treatments, it is assumed that no volume-change occurs in the association
reaction, i.e., the reaction is pressure-independent. However, a number of proteins are
found to exhibit a mostly positive volume change (ref. 64). In such cases, data obtained at
different initial concentrations or at different rotor-speeds will not fall on the single
M_w^{app} versus c curve. Further refinements of theory so as to take the pressure effects into
account are required at present. In order to minimize the effects experimentally, a low-
speed, conventional equilibrium run seems more suitable than a high-speed, meniscus
depletion run for study of reacting systems, as it is for ordinary molecular weight
determinations.

Table 3. Sedimentation analysis of Limulus haemocyanin (36 S component)[*]

Parameter	Value	Confidence region	Units
M_1 x 10-3	847	(837, 861)	
K_2	41.8	(38.9, 44.9)	$1\ g^{-1}$
K_4	69.5	(29.5, 106.2)	$1^3 g^{-3}$

For experimental conditions, see the text. The monomer under these
conditions is the 24 S component. The major species is the dimer, the
36 S component. The 60 S component is the tetramer, having an M value
of 3.39 million. Ninety-five per cent confidence region is given in
parentheses. From Johnson and Yphantis (ref. 63).
*Sedimentation coefficients are expressed in Svedberg units (S).

REFERENCES

1. T. Svedberg and H. Rinde, J. Am. Chem. Soc., 46, 2677 (1924).
2. T. Svedberg, Z. phys. Chem., 121, 65 (1926).
3. T. Svedberg and R. Fåhraeus, J. Am. Chem. Soc., 48, 430 (1926).
4. T. Svedberg and J.B. Nichols, J. Am. Chem. Soc., 48, 3081 (1926).
5. W. Weaver, Phys. Rev., 27, 499 (1926).
6. A. Tiselius, Z. phys. Chem., 124, 449 (1926).
7. T. Svedberg and A. Lysholm, Nova Acta R. Soc. Upsal., vol. extra ord. (1927).
8. T. Svedberg and J.B. Nichols, J. Am. Chem. Soc., 49, 2920 (1927).
9. T. Svedberg and E. Chirnoaga, J. Am. Chem. Soc., 50, 1399 (1928).
10. K.O. Pedersen, Z. phys. Chem., A 170, 41 (1934).
11. W.D. Lansing and E.O. Kraemer, J. Am. Chem. Soc., 57, 1369 (1935).
12. M. Meselson and F.W. Stahl, Proc. Natl. Acad. Sci. U.S.A., 44, 671 (1958).
13. J.E. Hearst and C.W. Schmid, in Methods in Enzymology (C.H.W. Hirs and S.N. Timasheff,
    ed.), vol. 27, p.111. Academic Press, New York (1973).
14. P.G. Squire and C.H. Li, J. Am. Chem. Soc., 83, 3521 (1961).
15. Th. G. Scholte, J. Polymer Sci. A-2, 8, 841 (1970).
16. W.G. McMillan and J.E. Mayer, J. Chem. Phys., 13, 276 (1945).
17. B.H. Zimm, J. Chem. Phys., 14, 164 (1946).
18. J.G. Kirkwood and F.P. Buff, J. Chem. Phys., 19, 774 (1951).

19. M. Gordon, C.G. Leonis and H. Suzuki, Proc. Roy. Soc. London, A **345**, 207 (1975).
20. C.G. Leonis, H. Suzuki and M. Gordon, Makromol. Chem., **178**, 2867 (1977).
21. H. Suzuki, Bull. Inst. Chem. Res., Kyoto Univ., **56**, 89 (1978).
22. T. Svedberg, Kolloid-z., Zsigmondy-Festschrift, Erg. Bd. zu **36**, 53 (1925).
23. R.J. Goldberg, J. Phys. Chem., **57**, 194 (1952).
24. H. Fujita, J. Chem. Phys., **63**, 1326 (1959).
25. J.W. Williams, K.E. Van Holde, R.L. Baldwin and H. Fujita, Chem. Rev., **58**, 715 (1958).
26. R.C. Deonier and J.W. Williams, Proc. Natl. Acad. Sci. U.S.A., **64**, 828 (1969).
27. H. Fujita, J. Phys. Chem., **73**, 1759 (1969).
28. H. Osterhoudt and J.W. Williams, J. Phys. Chem., **69**, 1050 (1965).
29. D.A. Albright and J.W. Williams, J. Phys. Chem., **71**, 2780 (1967).
30. H. Utiyama, N. Tagata and M. Kurata, J. Phys. Chem., **73**, 1448 (1969).
31. T. Kotaka, N. Donkai and H. Inagaki, J. Polymer Sci. A-2, **9**, 1379 (1971).
32. K.O. Pedersen, in The Ultracentrifuge (by T. Svedberg and K.O. Pedersen), p.304. Oxford University Press (1940).
33. D.A. Yphantis, Biochemistry, **3**, 297 (1964).
34. H. Suzuki, Brit. Polymer J., **11**, 211 (1979).
35. K.E. Van Holde and J.W. Williams, J. Polymer Sci., **11**, 243 (1953).
36. Th. G. Scholte, Paper read at the IUPAC Symposium, Helsinki, 1972.
37. F.C. Bancroft and D. Freifelder, J. Mol. Biol., **54**, 537 (1970).
38. C.W. Schmid and J.E. Hearst, Biopolymers, **10**, 1901 (1971).
39. S.B. Dubin, in Methods in Enzymology (C.H.W. Hirs and S.N. Timasheff, ed.), vol. **26**, p.119. Academic Press, New York (1972).
40. S.B. Dubin, G.B. Benedek, F.D. Bancroft and D. Freifelder, J. Mol. Biol., **54**, 547, (1970).
41. I. Rubenstein and S.B. Leighton, Biophys. Soc. (U.S.) Abstr., 209a (1971).
42. B.H. Zimm, Biophys Chem., **1**, 274 (1974).
43. O. Lamm, Arkiv Kemi, Mineral., Geol., **17A**, No. 25 (1944).
44. See, for example, E.F. Casassa and H. Eisenberg, Adv. Protein Chem., **19**, 287 (1964).
45. D.E. Roark and D.A. Yphantis, Ann. N.Y. Acad. Sci., **164**, 245 (1969).
46. D.K. Hancock and J.W. Williams, Biochemistry, **8**, 2598 (1969).
47. D.C. Teller, in Methods in Enzymology (C.H.W. Hirs and S.N. Timasheff, ed.), vol. **27**, p.346. Academic Press, New York (1973).
48. K.E. Van Holde, in The Proteins, the third edition (H. Neurath and R.L. Hill, ed.), vol. **1**, p.225. Academic Press, New York (1975).
49. H. Kim, R.C. Deonier and J.W. Williams, Chem. Rev., **77**, 659 (1977).
50. E.T. Adams, Jr. and J.W. Williams, J. Am. Chem. Soc., **86**, 3454 (1964).
51. E.T. Adams, Jr., Biochemistry, **6**, 1864 (1967).
52. N.R. Langerman and I.M. Klotz, Biochemistry, **8**, 4746 (1969).
53. R.F. Steiner, Arch. Biochem. Biophys., **39**, 333 (1952); **49** 400 (1954).
54. K.E. Van Holde and G.P. Rossetti, Biochemistry, **6**, 2189 (1967).
55. H.G. Elias and R. Bareiss, Chimia, **21**, 53 (1967).
56. A.J. Sophianopoulos and K.E. Van Holde, J. Biol. Chem., **243**, 1804 (1964).
57. P.W. Chun and S.J. Kim, Biochemistry, **9**, 1957 (1970).
58. D.E. Roark and D.A. Yphantis, Biochemistry, **10**, 3241 (1971).
59. P.W. Chun, S.J. Kim, J.D. Williams, W.T. Cope, L.H. Tang and E.T. Adams, Jr., Biopolymers, **11**, 197 (1972).
60. H. Eisenberg and G. Tomkins, J. Mol. Biol., **31**, 37 (1968).
61. W.P. Reinhardt and P.G. Squire, Biochim. Biophys. Acta, **94**, 566 (1966).
62. M.L. Johnson, J.J. Correia, D.A. Yphantis and H.R. Halvorson, Biophys. J., **36**, 575 (1981).
63. M.L. Johnson and D.A. Yphantis, Biochemistry, **17**, 1448 (1978).
64. See, for example, W.F. Harrington and G. Kegels, in Methods in Enzymology (C.H.W. Hirs and S.N. Timasheff, ed.), vol. **27**, p.306. Academic Press, New York (1973).

16. Thermal Diffusion and Sedimentation

Francis J. Bonner

Vanderbilt University, Nashville, Tennessee 37235, USA

Abstract — Both analytical ultracentrifugation and thermal diffusion can
be viewed as physical situations where a unidirectional force field is
superimposed on a system held at steady state, resulting in similar,
useful experimental arrangements and mathematical relations. This
present review emphasises these similarities and also discusses several
noteworthy features in the state-of-the-art of macromolecular thermal
diffusion, viz. discrepancies between thermogravitational columns and
"convectionless" cells, complications arising from residual convection
in "convectionless" cells, the utility of an expression for the thermal
force based on elastic, high-frequency, thermal waves, peculiar
phenomena near solubility limits indicating sharp fractionations and
relatively high thermal diffusion co-efficients, conflicting views of
behaviour, near critical points of mixing, and some biological
implications, especially for assemblies of cells the size of tumours and
organs. Equations are also derived that relate molecular size and shape
to the thermal diffusion coefficient and other measurable parameters.

Thermal diffusion becomes a logical extension of Svedberg´s interests in ultracentrifugation
when it is viewed as another situation where a force field is superimposed on a system held
at steady state. Even an appreciation of its biological implications is present in at least
one of his publications (ref. 1). Thus, it should not be surprising that thermal diffusion
in liquids, especially of macromolecules, has been a subject of research activity over an
extended period at the Institute of Physical Chemistry in Uppsala.

Attempts to relate thermal diffusion in liquids to fundamental and measurable system
properties and parameters have generally been based on a statistical treatment of Brownian
motion, theories of the liquid state, and the principles of irreversible thermodynamics
(refs. 2-5). Unfortunately, the derived relations are generally complex, with individual
terms inaccessible to direct evaluation and lacking a single, precisely defined physical
meaning. These relations have little practical value in predicting the magnitude of the
thermal diffusion or even its directions, i.e. the exceptions to the usual migration of
solute away from higher temperatures toward lower temperatures. More recently, however, a
relatively simple and potentially very useful equation relating the force on a particle (or
a molecule), caused by a thermal field, to measurable system properties and parameters has
been derived from the concept that heat is transported in the form of high frequency elastic
waves (ref. 5), viz.

$$\underline{F}_{t,p} = -2\gamma_{1,p}^2 \underline{K}_1 \underline{A}_p \{\frac{1}{\omega_1} - \frac{1}{\omega_p}\}(\frac{d\underline{T}}{d\underline{y}})_1 \tag{1}$$

where $\underline{F}_p$ is the thermal field force on a particle; γ, the transmission coefficient; $\underline{K}$, the
thermal conductivity; $\underline{A}$, cross-sectional area; ω, the wave propagation velocity; $d\underline{T}/d\underline{y}$, the
applied temperature gradient; and the subscripts 1, p, and t stand for liquid dispersing
medium, particle, and thermal respectively. Gaeta (ref. 5) has shown that this expression
can predict the just mentioned reversals in the usual direction for thermal diffusion.

The relative orientation of thermal diffusion to sedimentation can have pronounced
effects. At one extreme are the thermogravitational columns, e.g. Clusius - Dickel type
columns (ref. 6). These are vertical columns with a temperature difference superimposed
horizontally across a narrow gap that leads into top and bottom reservoirs (or
compartments). The functioning of such columns is generally described as an interplay
between horizontal thermal diffusion and vertical convection downward along the cold wall
and upward along the hot wall. At the other extreme are cells in which a temperature
difference is superimposed vertically across a column of liquid, e.g. Tanner (ref. 7) and
Longsworth (ref. 8) type cells. In such cells the directions for sedimentation and for
thermal diffusion are parallel.

The publication by Debye and Bueche in 1948 (ref. 9) on the thermogravitational
fractionation of polystyrene not only showed that such columns could be used to fractionate
polymers but also concluded that the thermal diffusion coefficient for polymers was strongly

affected by molecular weight and concentration. This work appears to have encouraged many subsequent studies using thermogravitational columns and Tanner/Longsworth type cells, many of which supported the findings of Debye and Bueche. Tyrrell (ref. 4)) provides a rather thorough review of the literature on thermal diffusion in liquids, including the phenomenological theory of thermogravitational columns, through the 1950´s.

Tanner/Longsworth type cells have often been erroneously considered convectionless. The discrepancies in the literature resulting from this as well as from not properly allowing for the dependence of the Fickian diffusion coefficient on concentration have been reviewed both by Rauch and Meyerhoff and by Norberg and Claesson (refs. 10-12).

In disagreement with conclusions based on thermogravitational experiments, investigators who have attempted to minimize convection and then to correct for residual convection as well as indirect concentration effects in Tanner/Longsworth cells (refs. 10-13) have found at most only a slight dependence of the thermal diffusion coefficient on polymer molecular weight and concentration. Eqn (1) can also be used to predict a similar slight dependence (ref. 5). This pervasive disagreement can again be seen in a relatively recent study (ref. 14) that repeated in a Longsworth type cell a thermogravitational study made on dextrans (ref. 15). These observations as well as the dissenting conclusions of Rauch and Meyerhoff (ref. 10) and of Ham (ref. 16) regarding the functioning of thermogravitational columns, viz. the dependence of the normal Fickian diffusion coefficient on concentration being more important than thermal diffusion, provide reasons for attributing fractionation and directional inversion (ref. 15) in thermogravitational columns to factors other than thermal diffusion and to reinterpret the experimental results previously thought to indicate a strong dependence of the thermal diffusion coefficient on molecular weight and concentration. Using a Longsworth type cell, Bonner (ref. 17), however, did find under restricted conditions the formation of sudden shifts in interference patterns indicating sharp fractionations and, correspondingly, relatively large thermal diffusion coefficients which were strongly dependent on molecular weight and concentration. This only occurred under conditions very close to precipitation where intermolecular aggregation could be expected. Eqn (1) could directly predict such behaviour in terms of an increasing cross-sectional area and also possibly somewhat more subtly by appropriate changes in the coupling coefficient and the term inside the bracket.

Two types of thermal diffusion measurements are carried-out in Tanner/Longsworth cells, viz. moving boundary and Soret equilibrium. Typically, a vertical column of liquid is held between an upper hot plate and a lower cold plate. In a moving boundary experiment this column is initially separated into solvent above and dense solution below by a sharp interface, which then spreads by Fickian diffusion and moves (usually toward the cold plate) by thermal diffusion. In a Soret equilibrium experiment the interplay of Fickian and of thermal diffusion establishes a steady state concentration gradient in an initially uniform column of solution. The following operational relations are generally applied:

A. For the moving boundary:

$$\underline{J} = \underline{J}_f + \underline{J}_t = -\underline{D}_f \frac{d\underline{c}}{d\underline{y}} - \underline{D}_t \underline{c} \frac{d\underline{T}}{d\underline{y}} \tag{2}$$

and therefore for the region of essentially constant concentration $\underline{c}_c$ below the interface

$$\underline{J} = \underline{J}_t = -\underline{D}_t \underline{c}_c \frac{d\underline{T}}{d\underline{y}} = \underline{v}_c \, \underline{c}_c, \tag{3}$$

consequently,

$$\underline{D}_t = -\frac{\underline{v}_c}{\tau}, \quad \text{where} \quad = \frac{d\underline{T}}{d\underline{y}}, \tag{4}$$

B. For Soret equilibrium:

$$\underline{J} = 0; \quad \therefore \, \underline{J}_t = -\underline{J}_f; \quad \therefore \, \underline{D}_t \underline{c} \tau = -\underline{D}_f \frac{d\underline{c}}{d\underline{y}}$$

and therefore,

$$\sigma_s = \frac{\underline{D}_t}{\underline{D}_f} = -\frac{1}{\tau \underline{c}} \frac{d\underline{c}}{d\underline{y}} = -\frac{1}{\tau} \frac{d\ln \underline{c}}{d\underline{y}} \tag{5}$$

consequently,

$$\ln \frac{\underline{c}_2}{\underline{c}_1} = -\sigma_s \tau (\underline{y}_2 - \underline{y}_1) \tag{6}$$

or

$$\frac{\underline{c}_2}{\underline{c}_1} = e^{-\sigma_s \tau (\underline{y}_2 - \underline{y}_1)}. \tag{7}$$

In the preceeding expressions,

J = flux of solute, D = diffusion coefficient,

c = concentration, T = temperature,

y = vertical position, v = velocity,

τ = temperature gradient,

σ_s = Soret coefficient and the subscripts f, t, c, l, stand for Fickian, thermal, constant, and liquid, respectively.

Analogous to the derivation of the Svedberg equation in ultracentrifugation, one can write a force balance on a particle (or a molecule) at a terminal velocity v_c in a vertical thermal field to be

$$F_{t,p} = fv_c - (\frac{M}{N_o})(1 - \rho \bar{V})g. \tag{8}$$

if

$$F_{t,p} \gg (\frac{M}{N_o})(1 - \rho \bar{V})g,$$

this reduces to

$$F_{t,p} = v_c, \tag{9}$$

which can be converted via eqn (3) to

$$F_{t,p} = - fD_t (\frac{dT}{dy})_l. \tag{10}$$

Substituting for $F_{t,p}$ in terms of eqn (1), then gives

$$D_t = 2f^{-1} \gamma^2_{1,p} K_1 A_p (\frac{1}{\omega_1} - \frac{1}{\omega_p}) \tag{11}$$

or alternatively, substituting for f via the Einstein relation

$$D_f = \frac{kT}{f} \quad \text{with} \quad k = \frac{R}{N_o} \quad \text{gives}$$

$$D_t = 2 \frac{D_f N_o}{RT} \gamma^2_{1,p} K_1 A_p (\frac{1}{\omega_1} - \frac{1}{\omega_p}) \tag{12}$$

as the desired expression relating the thermal diffusion coefficient to determinable system parameters. Assuming a molecular shape, the molecular weight can be introduced into eqn (12) via the cross-sectional area. If a spherical shape is assumed,

$A_p = \pi r^2_p$, $V_p = \frac{4}{3}\pi r^3_p$, and $V_p = \frac{M}{N_o}\bar{V}_p$. It then follows directly that

$$M_p = \left[\frac{\beta}{\gamma^2 K_1 (\frac{1}{\omega_1} - \frac{1}{\omega_p})} \right]^{3/2} \left[\frac{RTD_t}{D_f} \right]^{3/2} \bar{V}^{-1}_p \tag{13}$$

where $\beta = 0.414\, N_o^{-\frac{1}{3}}$, N_o = Avogadro's number, M = molecular wt., R = universal gas constant, r = radius, V = volume, $\bar{V}$ = partial specific volume, k = Boltzman's constant, f = friction factor. Table 1 emphasises the similarity between analytical ultracentrifugation and thermal diffusion. The thermal diffusion coefficient can be considered a mobility in the same sense as the sedimentation coefficient. Perhaps it should have been called a thermal sedimentation coefficient!

Table 1   ANALYTICAL ULTRACENTRIFUGATION VS THERMAL DIFFUSION

SEDIMENTATION COEFFICIENT:	THERMAL DIFFUSION COEFFICIENT

$$\underline{s} = \frac{v}{\omega^2 r}$$

$$\underline{D}_t = -\frac{v}{\tau}$$

VELOCITY SEDIMENTATION:
(SVEDBERG EQUATION)

$$\underline{M} = \frac{sRT}{\underline{D}_f \;(1 - \overline{V}\rho)}$$

MOVING BOUNDARY

$$\underline{M} = \left[\frac{\beta}{\gamma^2 \underline{K}(\frac{1}{\omega_1} - \frac{1}{\omega_p})}\right]^{3/2} \left[\frac{RTD_t}{\underline{D}_f}\right]^{3/2} \left[\frac{1}{\overline{V}_p}\right]$$

where $\beta = 0.414\ \underline{N}_o^{-1/3}$

EQUILIBRIUM SEDIMENTATION

$$\underline{M} = \frac{2\underline{RT}\{\ln(\underline{c}_2/\underline{c}_1)\}}{\omega^2\;(1 - \overline{V}\rho)(\underline{r}_2^2 - \underline{r}_1^2)}$$

SORET EQUILIBRIUM

$$\underline{M} = \left[\frac{\beta}{\gamma^2 \underline{K}(\frac{1}{\omega_1} - \frac{1}{\omega_p})}\right]^{3/2} \left[\frac{RT\{\ln(\underline{c}_2/\underline{c}_1)\}}{-\tau(\underline{y}_2 - \underline{y}_1)}\right]^{3/2} \left[\frac{1}{\overline{V}_p}\right]$$

Note: ω = angular velocity

Note: ω = wave propagation velocity.

Descriptions of the design and operation of the Uppsala thermal diffusion cell are available (refs. 11-13). Considerable effort was expended in both design and operation, which included locating the cell in a heavily insulated basement room on an extra heavy optical bench firmly attached to a massive concrete foundation. Nevertheless, shifts in the interference pattern relative to fixed interference marks just beyond the top and bottom of the liquid column indicated that convection could at times account for around 20 - 40% of the apparent value of the thermal diffusion coefficient (ref. 13,17). Fig. 1 suggests the kinds of convection and the resultant changes in a plot of fringe function (ref. 13) versus vertical position that could occur. An upward shift in the solvent plateau above the interface in a moving boundary experiment could result from convection in the upper part of the cell planing-off the interface which in turn would increase the apparent magnitude of the thermal diffusion coefficient as described by Norberg and Claesson (ref. 12). An upward shift of the solvent plateau combined with a downward shift of the solution plateau could be attributed to convective interchange between the regions above and below the interface along with convective mixing within these regions. Convective transport of solute from near the cold plate to near the hot plate without sufficient mixing above and below the interface could result in the loss of the plateaus. All these various changes in the fringe function curves have been observed (ref. 18). These convections do not appear traceable to purturbations caused by the formation of the interface (ref. 13). They might simply be caused by the surroundings in terms of heat exchange and vibrations but they might also be the consequence of the thermal diffusion itself, viz. the countercurrent flows of hot and cold mass that must thermally re-equilibrate with their immediate surroundings. In keeping with this possibility, convection does appear to increase along with increases in thermal diffusion in these moving boundary experiments (ref. 18). Such associated convection might contribute to the sharp, relatively rapidly moving secondary interfaces that formed in situ when the moving boundary experiments were carried-out close to precipitation of solute (ref. 17). Similar secondary interfaces have also been formed under such conditions in Soret equilibrium experiments (ref. 18).

As a general mechanism for active transport, i.e. mass transport up a concentration gradient, thermal diffusion becomes an intriguing possibility for mass transport in living organisms. Because of their thinness, an immeasurably small temperature difference, e.g. $0.0001^{\circ}C$, across bimolecular lipid membranes could produce temperature gradients much greater than those required in moving boundary thermal diffusion measurements. A recent assessment of thermal diffusion as a mechanism for biological transport concluded that ensembles of cells the size of tumours or organs could produce particularly favourable conditions for biological transport (ref. 19).

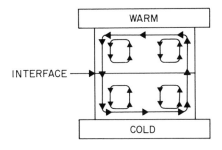

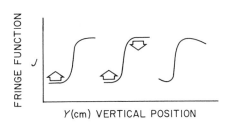

Fig. 1.

On both theoretical and experimental grounds, Fickian diffusion coefficients are expected to tend to zero at critical points of mixing (refs. 3, 20-24). The behaviour of thermal diffusion coefficients is more controversial. There had been no adequate theoretical basis for predicting behaviour. Some experiments indicated a substantial increase (refs. 20, 25, 26) while others indicated little change (refs. 21, 27). The already mentioned moving boundary experiments near the solubility limits of polymers (ref. 17) suggest that thermal diffusion increases as molecular association increases, which is in keeping with the theoretical basis provided by eqn (1). An increasing Soret coefficient requires only that the thermal diffusion coefficient decrease toward zero slower than the Fickian diffusion coefficient does, since by definition the Soret coefficient is simply the ratio of the thermal diffusion coefficient to the Fickian diffusion coefficient. Obviously, an increasing thermal diffusion coefficient would provide a significantly more promising condition for active transport since the thermal diffusive flux would be increasing and the potential for active transport, i.e. the limiting concentration difference expressed by the Soret coefficient eqn (7), would be increasing more rapidly.

REFERENCES

1. N. Gralen and T. Svedberg, Naturwissenshaften **29**, 270 (1941).
2. I. Prigogine, Introduction to Thermodynamics of Irreversible Processes, (Charles C. Thomas, ed.) Springfield, Illinois (1955).
3. R. Haase, Thermodynamics of Irreversible Processes, Addison-Wesley Pub. Co., Reading, Mass (1969), translation of R. Haase, Thermodynamik der irreverisblen Prozesse, Dr. Dietrick Steinkopf Verlag, Darmstadt (1963).
4. H.J.V. Tyrrell, Diffusion and Heat Flow in Liquids, Butterworths, London (1961).
5. F.S. Gaeta, Fourth Winter School of Biophysics of Membrane Transport, p. 68, Poland (1977).
6. K. Clusius and G. Dickel, Naturwissenschaften **26**, 546 (1938) and **27**, 148 (1939).
7. C.C. Tanner, Trans. Faraday Soc., **23**, 75 (1927) and **49**, 611 (1953).
8. L.G. Longsworth, J. Phys. Chem. **61**, 1557 (1957).
9. P. Debye and A.M. Bueche, Thermal Diffusion of Polymer Solutions (in High Polymer Physics, a Symposium, H.A. Robinson, ed.), p. 497. Remsen Press Div., Chemical Publishing, New York (1948).
10. B. Rauch and G. Meyerhoff, J. Phys. Chem. **67**, 946 (1963).
11. P.H. Norberg, Abhandle. Deut. Akad. Wiss., Berlin, Kl. Med. No. 6, II. Jenaer Symposium 1 (1964).
12. P.H. Norberg and S. Claesson, ACTA IMEKO **4**, 501 (1964).
13. F.J. Bonner, Arkiv Kemi **27**, 97, 115 (1967).
14. F.J. Bonner, Chemica Scripta **3**, 149 (1973).
15. F.S. Gaeta, A. Di Chiara and G. Perna, Il Nuovo Cimento, LXVIB, N. 2, 260 (1970).
16. J.S. Ham, J. Appl. Phys. **31**, 1853 (1960).
17. F.J. Bonner, Arkiv Kemi **27**, 19, 129, 139 (1967).
18. F.J. Bonner, unpublished work.
19. F.J. Bonner and L.-O. Sundelöf, Z. Naturforsch. **39C**, 656 (1984).
20. S. Claesson and L.-O. Sundelöf, J. Chim. Phys. **54**, 914 (1957).
21. M. Giglio and A. Vendramini, Phys. Rev. Letters **34**, 561 (1975).
22. R. Haase, Ber. Bunsenges. Physik. Chem. **76**, 256 (1972).
23. R. Bergman and L.-O. Sundelöf, Eur. Polymer J. **13**, 881 (1977).
24. L.-O. Sundelöf, Ber. Bunsenges. Phys. Chem. **83**, 329 (1979).
25. G. Thomaes, J. Chem Phys. 25, 32 (1956).
26. L.-O. Sundelöf, Chem. Zvesti 25, 203 (1971).
27. L.-P. Sundelöf, et al., IUPAC Symposium Macromolecules (Helsinki) Preprint **3**, 129 (1972).

V Macromolecular Solutions and Gels

17. Light Scattering from Mixtures of Homopolymers and Copolymers. Theoretical Results and Experimental Examples

Henri Benoit, Mustapha Benmouna*, Claude Strazielle and Carlos Cesteros**

Centre de Recherches sur les Macromolécules (CNRS) 6, rue Boussingault
67083 Strasbourg Cedex, France

* Université de Tlemcen, TLEMCEN, Algérie

** Université du Pays Basque, BILBAO, Espagne.

Abstract - A general equation giving the light (or the neutrons) scattered by a solution of homopolymers and copolymers at any concentration is presented. Some of its applications to results obtained by light scattering are presented.

I. THEORETICAL

INTRODUCTION

Light and neutron scattering techniques have become major tools for studying physico-chemical properties of polymer solutions. In this paper, we would like to show how it is possible to write a general equation giving the intensity of light or neutrons scattered at any angle and any concentration for a solution containing polymers and copolymers, mono-disperse or polydisperse, and to show a few examples of its applications. The elastic scattering intensity is generally the sum of two quantities: the first one is the intra-molecular form factor denoted by $P(q)$ and the second one, $Q(q)$ is the intermolecular or total correlation function. For a solution of n identical homopolymers per unit volume, we write the scattering intensity as:

$$I(q) = K\ nN^2[\ P(q) + Q(q)] \tag{1}$$

where K is a constant depending on the type of radiation (light or neutrons), $q = (4\pi/\lambda)$ $\sin(\theta/2)$, λ is the wavelength of the incident of the incident radiation, θ the scattering angle, and N is the degree of polymerization. $Q(q)$ is defined as:

$$Q(q) = \underset{\ell}{\Sigma}\ \underset{m}{\Sigma}\ \underset{j}{\Sigma}\ \underset{k}{\Sigma} < \exp(i q \cdot R_{\ell jmk}) > \tag{2}$$

where $R_{\ell jmk}$ is the vector connecting monomer j on chain ℓ to monomer k on chain m. The cornered brackets designate the thermal average with respect to the equilibrium distribution function of the vector $R_{\ell jmk}$.

P(q) is known for various shapes of macromolecules such as a Gaussian chains (Debye function), hard spheres, or rigid rods. The main theoretical difficulty is to calculate $Q(q)$. This calculation is the object of the first part of this paper. In preceding papers (refs. 1 - 3) we obtained equations for $Q(q)$ by summing infinite series of diagrams. Here, we would like to use directly a generalization of the Ornstein-Zernike (ref. 4) theory which gives the same results in a straightforward manner. These results are from our point of view very useful for interpreting scattering experiments on complex systems since they allow intramolecular contributions to be extracted from the total scattering signals (ref. 5, 6).

THE CASE OF IDENTICAL HOMOPOLYMERS

We consider a solution made of n identical homopolymers (per unit volume) embedded in a solvent of arbitrary quality (good or theta). The OZ, equation for this system can be written as follows

$$H(q) = C(q) + H(Q)\ C(q) \tag{3}$$

119

where in the q-space $\underline{C}(\underline{q})$ is the direct correlation function between a pair of chains and $\underline{H}(\underline{q}$ is the indirect correlation function which is related to $\underline{Q}(\underline{q})$ by:

$$\underline{H}(\underline{q}) = \underline{Q}(\underline{q})/\underline{P}(\underline{q}) \tag{4}$$

$\underline{C}(\underline{q})$ must be specified in order to calculate $\underline{H}(\underline{q})$. In simple liquids theory, there are several classical models which give equations relating $\underline{C}(\underline{q})$, $\underline{H}(\underline{q})$ and the potential of pair interactions. Depending on the level of accuracy desired and the particular problem at hand, one uses the so-called Percus-Yevick (PY), the Hypernetted-chain (HNC) or some other scheme (ref. 7). These models correspond to series of diagrams representing the interactions between particles, which are truncated at different levels. Here we shall use the so-called single contact approximation and write:

$$\underline{C}(\underline{q}) = - \underline{n}\,\underline{v}\,\underline{N}^2\,\underline{P}(\underline{q}) \tag{5}$$

where $\underline{v}$ is the excluded volume parameter for a pair of monomers:

$$\underline{v} = - \int \underline{d}^3\underline{R}\,[\underline{g}(\underline{R})-1\,]\,e^{i\,\underline{q}\,\underline{R}} \tag{6}$$

and $\underline{g}(\underline{R})$ is the pair distribution function. To obtain eqn (5) it is sufficient to assume that the interactions between monomers are limited to distances short as compared to $\underline{q}-1$, writing:

$$\underline{g}(\underline{R}) - 1 = -\,\underline{v}\,\delta(\underline{R}) \tag{7}$$

where δ is the DIRAC delta function. This approximation can be relaxed if the range of interaction is finite and this can lead to a $\underline{q}$- dependent excluded volume parameter $\underline{v}(\underline{q})$. Solving eqn (3) for $\underline{H}(\underline{q})$ gives:

$$\underline{H}(\underline{q}) = \frac{\underline{C}(\underline{q})}{I - \underline{C}(\underline{q})} \tag{8}$$

and using eqn (5):

$$\underline{H}(\underline{q}) = -\frac{\underline{v}\,\underline{N}^2\underline{P}(\underline{q})}{1+\,\underline{v}\underline{n}\underline{N}^2\underline{P}(\underline{q})} \tag{9}$$

Substituting this result into eqn (1), we obtain:

$$\underline{K}^{-1}\,\underline{I}(\underline{q}) = \frac{\underline{n}\underline{N}^2\underline{P}(\underline{q})}{1+\,\underline{v}\underline{n}\underline{N}^2\underline{P}(\underline{q})} \tag{10}$$

This result can be written in the reciprocal form using the standard notation (ref.8):

$$\frac{Kc}{\underline{I}(\underline{q})} = \frac{1}{\underline{M}\,\underline{P}(\underline{q})} + 2A_2\underline{c} \tag{11}$$

where $\underline{c}$ is the concentration in weight fraction, $\underline{M}$ the molecular weight, and A_2 the generalized second virial coefficient which includes all terms of higher orders in the concentration $\underline{c}$ ($2A_2\underline{M}\underline{c} = \underline{v}\underline{n}\underline{N}^2$). The $\underline{c}$-dependence of A_2 or $\underline{v}$ can be determined in the framework of the Flory-Huggins theory. One obtains easily:

$$\underline{v}(\underline{c}) = \underline{v}_s\,((1/\phi_s\,) - 2\chi) \tag{12}$$

where $\underline{v}_s$ is the volume occupied by a solvent molecule, ϕ_s is the volume fraction of the solvent and χ is the Flory's interaction parameter (ref.9).

THE CASE OF A MIXTURE OF HOMOPOLYMERS

The above formalism is now applied to a mixture of different homopolymers. For simplicity, let us consider a solution of two homopolymers denoted A and B. The scattering intensity is proportional to:

$$\underline{I}(\underline{q}) = \underline{a}^2\underline{S}_a(\underline{q}) + \underline{b}^2\,\underline{S}_b(\underline{q}) + 2\,\underline{a}\underline{b}\,\underline{S}_{ab}(\underline{q}) \tag{13}$$

where a and b are the excess scatterings for monomers A and B, respectively, and $\underline{S}_a(\underline{q})$, $\underline{S}_b(\underline{q})$ and, $\underline{S}_{ab}(\underline{q})$ are the partial structure factors:

$$\underline{S}_a(\underline{q}) = \underline{n}_a\underline{N}^2{}_a\,[\,\underline{P}_a(\underline{q}) + \underline{Q}_a(\underline{q})\,] \tag{14}$$

$$\underline{S}_b(\underline{q}) = \underline{n}_b\underline{N}^2{}_b \ [\ \underline{P}_b(\underline{q}) + \underline{Q}_b(\underline{q})\ ] \tag{15}$$

$$\underline{S}_{ab}(\underline{q}) = \sqrt{\underline{n}_a\underline{n}_b}\ \underline{N}_a\underline{N}_b\ \underline{Q}_{ab}(\underline{q}) \tag{16}$$

The OZ equations for mixtures of particles are known as (ref. 10):

$$\underline{H}_{\underline{\ell}\,\underline{m}}(\underline{q}) = \underline{C}_{\underline{\ell}\,\underline{m}}(\underline{q}) + \underset{\underline{n}}{\Sigma}\ \underline{H}_{\underline{\ell}\,\underline{n}}(\underline{q})\ \underline{C}_{\underline{n}\underline{m}}(\underline{q}) \tag{17}$$

where the indices $\underline{\ell}$, $\underline{m}$ and $\underline{n}$ run over the species A and B, $\underline{C}_{\underline{\ell}\,\underline{m}}(\underline{q})$ are the direct correlation functions and $\underline{H}_{\underline{\ell}\underline{m}}(\underline{q})$ are related to $\underline{Q}_{\underline{\ell}\underline{m}}(\underline{q})$ by

$$\underline{H}_{\underline{\ell}\underline{m}}(\underline{q}) = \underline{Q}_{\underline{\ell}\,\underline{m}}(\underline{q})\ /\ \sqrt{\underline{P}_{\underline{\ell}}(\underline{q})\ \underline{P}_{\underline{m}}(\underline{q})}$$

where $\underline{\ell}$ and $\underline{m}$ can be either $\underline{a}$ or $\underline{b}$. Solving eqn (17) for $\underline{H}_{aa}(\underline{q})$ and $\underline{H}_{ab}(\underline{q})$, one obtains:

$$\underline{H}_{aa}(\underline{q}) = \frac{\underline{C}_{aa} - (\underline{C}_{aa}\underline{C}_{bb} - \underline{C}^2{}_{ab})}{1 - \underline{C}_{aa} - \underline{C}_{bb} + (\underline{C}_{aa}\underline{C}_{bb} - \underline{C}^2{}_{ab})} \tag{18}$$

and

$$\underline{H}_{ab}(\underline{q}) = \frac{\underline{C}_{ab}}{1 - \underline{C}_{aa} - \underline{C}_{bb} + (\underline{C}_{aa}\underline{C}_{bb} - \underline{C}^2_{ab})} \tag{19}$$

$\underline{H}_{bb}(\underline{q})$ is deduced from $\underline{H}_{aa}(\underline{q})$ by interchanging the indices $\underline{a}$ and $\underline{b}$. The single contact approximation is again used to obtain the direct correlation functions:

$$\underline{C}_{\underline{\ell}\,\underline{m}} = -\ \underline{v}_{\underline{\ell}\underline{m}}\ \underline{N}_{\underline{\ell}}\ \underline{N}_{\underline{m}}\ \sqrt{\underline{n}_{\underline{\ell}}\ \underline{n}_{\underline{m}}\ \underline{P}_{\underline{\ell}}}\ (\underline{q})\underline{P}_{\underline{m}}(\underline{q}) \tag{20}$$

where $\underline{v}_a$, $\underline{v}_b$ and $\underline{v}_{ab}$ are the excluded volume parameters relative to pairs of monomers AA, BB and AB, respectively.

Using Flory-Huggins theory as in the case of eqn (12), one obtains:

$$\underline{v}_a = \underline{v}_s\ (\frac{1}{\phi_s} - 2\chi_{as}) \tag{21}$$

$$\underline{v}_b = \underline{v}_s\ (\frac{1}{\phi_s} - 2\chi_{bs}) \tag{22}$$

$$\underline{v}_{ab} = \underline{v}_s\ (\frac{1}{\phi_s} - \chi_{as} - \chi_{bs} + \chi_{ab}) \tag{23}$$

where χ_{as}, χ_{bs}, and χ_{ab} are the Flory's interaction parameters.

Combining eqns (13) through (20) one obtains the scattering intensity as:

$$\underline{I}(\underline{q}) = \frac{a^2\underline{x}_a + b^2\underline{x}_b + (a^2\underline{v}_b + b^2\underline{v}_a - 2ab\underline{v}_{ab})\underline{x}_a\underline{x}_b}{1 + \underline{v}_a\underline{x}_a + \underline{v}_b\underline{x}_b + (\underline{v}_a\underline{v}_b - \underline{v}^2{}_{ab})\underline{x}_a\underline{x}_b} \tag{24}$$

where $\underline{x}_a = \underline{n}_a\underline{N}^2{}_A\underline{P}_a(\underline{q})$ and $\underline{x}_b = \underline{n}_b\underline{N}^2{}_b\underline{P}_b(\underline{q})$. This result can be used to investigate many situations of practical interest, some of them will be discussed later in the experimental section. This result was obtained independently by Wu and Rawiso (ref. 11) using the Random Phase Approximation (RPA). This implies that this theory is valid within the limits of the RPA and hence, it cannot describe properly systems in which the concentration fluctuations are very strong (ref. 12). In spite of this shortcoming of the theory, one can still attempt a description of the conditions where the scattering intensity becomes infinite at $\underline{q}$ = 0; this gives the following spinodal equation:

$$1 + \underline{v}_a\underline{n}_a\underline{N}^2{}_a + \underline{v}_b\underline{n}_b\underline{N}^2{}_b + (\underline{v}_a\underline{v}_b - \underline{v}^2{}_{ab})\underline{n}_a\underline{n}_b\underline{N}^2{}_a\underline{N}^2{}_b = 0 \tag{25}$$

For given molecular weights, this equation is a relation between the relative composition and the interaction parameters for which a spinodal decomposition may occur. This problem is extensively discussed in the literature and a fair number of measurements mainly by light scattering have been reported and analysed (refs. 13 - 15) using eqn (25). It is worthwhile to note that there may exist a finite value $q*$ of the wave vector $\underline{q}$ at which the intensity becomes infinite, before reaching the classical spinodale:

$$1 + \underline{v}_a\underline{n}_a\underline{N}^2{}_a\underline{P}_a(\underline{q}*) + \underline{v}_b\underline{n}_b\underline{N}^2{}_b\underline{P}_b(\underline{q}*) + (\underline{v}_a\underline{v}_b - \underline{v}^2{}_{ab})\ \underline{n}_a\underline{n}_b\underline{N}^2{}_a\underline{N}^2{}_b\underline{P}_a(\underline{q}*)\underline{P}_b(\underline{q}*) = 0 \tag{26}$$

for $\underline{q}* \neq 0$. This problem has been discussed in refs. (4) and (16) for a system in the bulk state and in a concentrated solution.

THE CASE OF COPOLYMERS

Let us assume that we have a solution of $\underline{n}$ copolymers of type AB. The partial structure factors can be written as:

$$\underline{S}_{\underline{\ell}\,\underline{m}}(\underline{q}) = \underline{n}\ [\underline{x}_{\underline{\ell}\,\underline{m}} + \underline{Q}_{\underline{\ell}\,\underline{m}}]$$

where $\underline{\ell}$, $\underline{m} = \underline{a}$ and $\underline{b}$, $\underline{x}_{\underline{\ell}\,\underline{m}} = \underline{n}\underline{N}_{\underline{\ell}}\ \ \underline{N}_{\underline{m}}\ \underline{P}_{\underline{\ell}\,\underline{m}}(\underline{q})$, and $\underline{Q}_{\underline{\ell}\,\underline{m}}$ are the total interparticle correlation functions. The OZ equations for this system are somewhat more subtle in the sense that there exist many more interaction terms than in the case of a mixture. It can be written as follows:

$$\underline{Q}_{\underline{\ell}\underline{m}}(\underline{q}) = \underline{n}^{\Sigma}\ [\ \underline{x}_{\ \underline{\ell}\,\underline{n}}\underline{C}_{\underline{m}\underline{n}}(\underline{q}) + \underline{Q}_{\ \underline{\ell}\,\underline{n}}(\underline{q})\ \underline{C}_{\underline{m}\underline{n}}(\underline{q})] \qquad (28)$$

where $\underline{\ell}$, $\underline{m}$ and $\underline{n}$ run over the species A and B. The direct correlation functions are obtained in the single contact approximation as follows:

$$\underline{C}_{\underline{\ell}\,\underline{\ell}}(\underline{q}) = -\ \underline{n}\underline{v}_{\underline{\ell}\,\underline{\ell}}\ \underline{N}^2\ \underline{P}_{\underline{\ell}}\ (\underline{q}) \qquad\qquad (\underline{\ell} = \ \underline{a},\underline{b}) \qquad (29)$$

$$\underline{C}_{\underline{\ell}\underline{m}}(\underline{q}) = -\ \underline{n}\underline{v}_{\underline{\ell}}\,\underline{N}_{\underline{\ell}}\underline{N}_{\underline{m}}\underline{P}_{\underline{\ell}\underline{m}}(\underline{q}) \qquad\qquad (\underline{\ell} \neq \ \underline{m}) \qquad (30)$$

$\underline{P}_{\underline{\ell}\underline{m}}(\underline{q})$ is the intramolecular contribution to the form factor due to interferences between the different species $\underline{\ell}$ and $\underline{m}$ within the copolymer chain. Combining these results together, we obtain the total correlation function as follows:

$$-\ \underline{D}(\underline{q})\ \underline{Q}_{\underline{a}\underline{a}}(\underline{q}) = \underline{v}_{\underline{a}}\underline{x}_{\underline{a}}^2 + 2\underline{v}_{\underline{a}\underline{b}}\underline{x}_{\underline{a}}\underline{x}_{\underline{a}\underline{b}} + \underline{v}_{\underline{b}}\underline{x}_{\underline{a}\underline{b}}^2 + (\underline{v}_{\underline{a}}\underline{v}_{\underline{b}} - \underline{v}_{\underline{a}\underline{b}}^2)(\underline{x}_{\underline{a}}\underline{a}_{\underline{b}} - \underline{x}_{\underline{a}\underline{b}}^2) \qquad (31)$$

$$-\ \underline{D}(\underline{q})\ \underline{Q}_{\underline{a}\underline{b}}(\underline{q}) = \underline{v}_{\underline{a}\underline{b}}\underline{x}_{\underline{a}\underline{b}}^2 + \underline{v}_{\underline{a}}\underline{x}_{\underline{a}}\underline{x}_{\underline{a}\underline{b}} + \underline{v}_{\underline{b}}\underline{x}_{\underline{b}}\underline{x}_{\underline{a}\underline{b}} + \underline{v}_{\underline{a}\underline{b}}\underline{x}_{\underline{a}}\underline{x}_{\underline{b}} + (\underline{v}_{\underline{a}}\underline{v}_{\underline{b}} - \underline{v}_{\underline{a}\underline{b}}^2)\underline{x}_{\underline{a}\underline{b}}(\underline{x}_{\underline{a}}\underline{x}_{\underline{b}} - \underline{x}_{\underline{a}\underline{b}}^2) \qquad (32)$$

where $\underline{D}(\underline{q})$ is given by:

$$-\ \underline{D}(\underline{q}) = 1 + \underline{v}_{\underline{a}}\underline{x}_{\underline{a}} + \underline{v}_{\underline{b}}\underline{x}_{\underline{b}} + 2\underline{v}_{\underline{a}\underline{b}}\underline{x}_{\underline{a}\underline{b}} + (\underline{v}_{\underline{a}}\underline{v}_{\underline{b}} - \underline{v}_{\underline{a}\underline{b}}^2)(\underline{x}_{\underline{a}}\underline{x}_{\underline{b}} - \underline{x}_{\underline{a}\underline{b}}^2) \qquad (33)$$

The scattering intensity can be put in the form:

$$\underline{I}(\underline{q}) = \frac{a^2\underline{x}_{\underline{a}} + b^2\underline{x}_{\underline{b}} + (a^2\underline{v}_{\underline{b}} + b^2\underline{v}_{\underline{a}} - 2ab\underline{v}_{\underline{a}\underline{b}})(\underline{x}_{\underline{a}}\underline{x}_{\underline{b}} - \underline{x}_{\underline{a}\underline{b}}^2)}{\underline{D}(\underline{q})} \qquad (34)$$

The analogy between this form and eqn (24) relative to the mixture of homopolymers is obvious; the extra terms involving $\underline{x}_{\underline{a}\underline{b}}$ and appearing here are due to the copolymer nature of the chains. It is interesting to note that at $\underline{q}=0$, the denominator of eqn (34) becomes:

$$\underline{D}(0) = 1 + \underline{n}\ (\underline{v}_{\underline{a}}\underline{N}^2_{\underline{a}} + \underline{v}_{\underline{b}}\underline{N}^2_{\underline{b}} + 2\underline{v}_{\underline{a}\underline{b}}\underline{N}_{\underline{a}}\underline{N}_{\underline{b}}) \qquad (35)$$

This equation means that, if the solvent is good for both types of unit, the incompatibility never leads to a phase transition.

The OZ formulation for polymer solutions which we have developed in this section is a mean field theory which is equivalent to the Random Phase Approximation. In fact, using Flory's values for $\underline{v}_{\underline{\ell}\underline{m}}$, one obtains the bulk limit by letting $\phi_{\underline{s}} = 0$ and one gets the same results as the RPA technique. For instance, in the case of a two blocks copolymer with identical blocks, one obtains (ref. 5):

$$\underline{K}^{-1}\underline{I}(\underline{q}) = \frac{P_{\frac{1}{2}} - P_T}{1 - \underline{N}\chi\ (P_{\frac{1}{2}} - P_T)} \qquad (36)$$

where $\underline{P}$ is the block form factor, $\underline{P}_T$ is the total form factor and χ is the interaction parameter between A and B monomers. All these equations describe systems that are not subject to strong fluctuations in the concentration and hence, they are valid above the overlap concentration $\underline{c}*$. Furthermore, we assumed that the interactions between pairs of monomers are represented by DIRAC delta functions. We neglected a number of situations in which either the chains interact via several points, or form loops, or make contacts of higher orders... We argued in ref. (4) that such effects can be incorporated within the final results simply by applying a renormalization argument in terms of polymer concentration both to the chain form factors and to the interaction parameters. A more elaborate calculation to improve these points is certainly needed and the recent attempt by Joanny and coworkers (ref. 17) is perhaps the way to proceed.

The OZ equations written down for a mixture in eqn (17) can be applied for hard spheres as well (ref. 10). In this case, the direct correlation functions $\underline{C}_{\ell m}(\underline{q})$ can be obtained using the single contact approximation as follows:

$$\underline{C}_{\underline{\ell}\underline{\ell}}(\underline{q}) = -8 \underline{C}_{\underline{v}\underline{\ell}} \Phi(\underline{u}_{\underline{\ell}}) \qquad (\underline{\ell} = \underline{a}, \underline{b}) \tag{37}$$

$$\underline{C}_{\underline{\ell}\,\underline{m}}(\underline{q}) = -\sqrt{\underline{C}_{\underline{v}\underline{\ell}}\,\underline{C}_{\underline{v}\underline{m}}} \; (\sqrt{\gamma} + \frac{1}{\sqrt{\gamma}})^3 \; \Phi(\frac{\underline{u}_{\underline{\ell}} + \underline{u}_{\underline{m}}}{2}) \qquad (\underline{\ell} = \underline{m}) \tag{38}$$

where $\Phi(\underline{u}) = \dfrac{3(\sin \underline{u} - \underline{u} \cos \underline{u})}{\underline{u}^3}$, $\underline{C}_{\underline{v}\underline{\ell}} = \underline{n}_{\underline{\ell}} \dfrac{4}{3} (\dfrac{\underline{D}_{\underline{\ell}}}{2})^3$ is the volume fraction of spheres whose number per unit volume is $\underline{n}_{\underline{\ell}}$, and diameter $\underline{D}_{\underline{\ell}}$, $\underline{u}_{\underline{\ell}} = \underline{q}\underline{D}_{\underline{\ell}}$ and $\gamma = \underline{D}_{\underline{\ell}}/\underline{D}_{\underline{m}}$. We note that $\underline{C}_{\underline{\ell}\underline{\ell}}$ $(\underline{q})$ is proportional to $\sqrt{\underline{P}(\underline{q})}$ whereas for linear chains $\underline{C}(\underline{q})$ was proportional to $\underline{P}(\underline{q})$ in the dilute regime and constant in the bulk state $\underline{n}\underline{Q}(\underline{q}) = -\underline{P}(\underline{q})$. Another important remark to be made has to do with the effect of polydispersity. One observes that since all our results are expressed in terms of the form factor $\underline{P}(\underline{q})$, these results remain valid for a polydisperse system provided that one replaces $\underline{N}^2\underline{P}(\underline{q})$ by $\Sigma \; \underline{N}_{\underline{i}}^2 \; \underline{P}_{\underline{i}}(\underline{q})$ where $\underline{i}$ runs over all molecular species. For instance, a mixture of homopolymers and copolymers could as well be represented by eqn (34) introducing for $\underline{x}_{\underline{a}}$, $\underline{x}_{\underline{b}}$ and $\underline{x}_{\underline{ab}}$ the appropriate definitions. More precisely for a system containing $\underline{n}_{\underline{a}}$ homopolymers A, $\underline{n}_{\underline{b}}$ homopolymers B, and $\underline{n}$ copolymers AB characterized by $\underline{N}_{\underline{A}}'$, $\underline{N}_{\underline{B}}'$, $\underline{P}_{\underline{a}}'$, $\underline{P}_{\underline{b}}'$ and $\underline{P}_{\underline{ab}}'$, it is sufficient to write:

$$\underline{x}_{\underline{a}} = \underline{n}_{\underline{a}}\underline{N}_{\underline{a}}^2 \, \underline{P}_{\underline{a}} + \underline{n} \; \underline{N}_{\underline{a}}'^2 \, \underline{P}_{\underline{a}}' \qquad\qquad \underline{x}_{\underline{b}} = \underline{n}_{\underline{b}}\underline{N}_{\underline{b}}^2\underline{P}_{\underline{b}} + \underline{n} \; \underline{N}_{\underline{b}}'^2\underline{P}_{\underline{b}}'$$

$$\underline{x}_{\underline{ab}} = \underline{n} \; \underline{N}_{\underline{a}}' \underline{N}_{\underline{b}}' \underline{P}_{\underline{ab}}' \tag{39}$$

In the following section we present some experimental results obtained by light scattering and attempt their interpretation using the theory developed here.

II. EXPERIMENTAL

In the preceding part we have established general equations enabling the calculation of the intensity scattered by a mixture of homopolymers and copolymers at any concentration. There are limitations to the validity of the equations. First, they do not give any information on $P(\underline{q},\underline{c})$ and on the values of the thermodynamical quantities $\underline{v}_{\underline{ij}}$, second, they do not take into account the fact that chains can have multiple contacts and, third, it is a mean field theory which does not use the progress introduced by scaling theories and renormalization arguments. The first object of experiments should therefore be to check these equations and to see how they can explain the experimental results. This programme can only be realized by the use of neutron scattering which allows the separation of the $\underline{P}$ and $\underline{Q}$ parameters of eqn (1) and has not yet been undertaken. What we would like here is to examine a few experiments already made by light scattering on mixtures of homopolymers and show how, at least qualitatively, they can be interpreted in the framework of this theory.

In order to simplify the interpretation of the data one usually tries to use a solvent for which the index of refraction is the same for one of the polymers and the solvent. If we assume that this is the case for polymer $\underline{b}$, we let $\underline{b} = 0$ in eqn (24) obtaining after some algebra

$$\underline{K} \; \underline{I}^{-1}(\underline{q}) = \underline{x}_{\underline{a}}^{-1} + \frac{\underline{v}_{\underline{a}} + (\underline{v}_{\underline{a}}\underline{v}_{\underline{b}} - \underline{v}_{\underline{ab}}^2)\underline{x}_{\underline{b}}}{1 + \underline{v}_{\underline{b}}\underline{x}_{\underline{b}}} \tag{40}$$

One sees immediately that this form is different from the equation for one homopolymer in solution. Since the second term depends on $\underline{q}$ through $\underline{x}_{\underline{b}}$ (the non visible polymer) one will not obtain the classical Zimm plots. They can be obtained only if $\underline{v}_{ab} = 0$, i.e. when there are no interactions between polymers A and B since in the case:

$$\underline{K}\ \underline{I}^{-1}(\underline{q}) = \frac{1}{\underline{x}_a} + \underline{v}_a$$

but this is more an exception than a rule, since even if it is true at one concentration, it cannot stay correct at others since $\underline{v}_{ab}$ depends on concentration. Therefore, quite generally, one will have distorted Zimm plots; a few examples of them have been given in ref. 3 and 4. Here we would like to study in more detail the following case: we assume that we have two polymers of different nature but otherwise identical. We mix them in equal proportion $\phi_a = \phi_b$ obtaining $\underline{x}_a$. Moreover we assume that the solvent is a "good" solvent for both writing $\underline{v}_a = \underline{v}_b$. Eqn (40) simplifies to:

$$\underline{K}\ \underline{I}^{-1}(\underline{q}) = \underline{x}^{-1} + \underline{v} - \frac{\underline{v}^2{}_{ab}\underline{x}}{1 + \underline{vx}} \tag{41}$$

We assume that the polymers are slightly incompatible. This means that we have a χ_{AB} parameter positive but small. From eqn (23) we obtain

$$\underline{v}_{ab} = \underline{v} + \underline{v}_s\,\chi_{AB}$$

and we can consider $\underline{v}_s\chi_{AB}$ as small compared to $\underline{v}$. The maximum concentration at the spinodale when the scattering intensity is infinite is obtained approximately as: $\underline{x}\ \underline{v}\underline{x}\,AB = 1$ or in terms of second virial coefficient

$$2\underline{A}_2\underline{Mc}\ [\chi_{AB}/(\frac{1}{\phi_s} - 2\chi_{As})] = 1$$

The zero angle scattering curve $\underline{c}$ versus $\underline{I}^{-1}(\underline{c})$ starts therefore with a positive slope, goes through a maximum and reach zero for the value of $\underline{c}$ given by the preceding equation as already observed in ref. 18. It is interesting also to look at the initial slope of the $\underline{c} =$ cste curves. For this purpose, we shall assume that $\underline{P}(\underline{q})$ is independent of $\underline{c}$ and evaluate the second term of the expansion of $\underline{c}\ \underline{I}^{-1}(\underline{q})$ as a function of $\underline{q}$ writing

$$\underline{N}^2\underline{nK}\ \underline{I}^{-1}(\underline{q}) = \underline{A} + \frac{q^2}{3}\ \underline{B}\ \underline{R}^2$$

calling $\underline{R}$ the radius of gyration of the polymers. One obtains for $\underline{B}$ the expression

$$\underline{B} = 1 + \frac{\underline{v}^2{}_{ab}\underline{n}^2\underline{N}^4}{(1 + \underline{n}\ \underline{N}^2\underline{v})^2} \tag{42}$$

When $\underline{n}$ increases, $\underline{B}$ becomes equivalent to $1 + \underline{v}^2{}_{ab}/\underline{v}^2$ and reaches two when ϕ goes to zero. In other words the slope increases from 1 to 2 when the concentration increases. In order to check these results, experiments have been made on a mixture of polystyrene and methyl-methacrylate. The molecular weight of these polymers were 1.4 and 1.3 millions and the polydispersity of the order of $\underline{M}_w/\underline{M}_n = 1.3$. The second virial coefficient for the polystyrene measured separately was $\underline{A}_2 = 2.7 \times 10^{-4}$. We used solutions made of equal amounts of PS and PMMA working in the range 0^2 to 0.006 g ml^{-1}. The results are represented on the Fig. 1 with theoretical curves for $\underline{q} = 0$. Qualitatively the agreement is good but quantitatively there are strong discrepancies. It is impossible to obtain from the theory a maximum as high as that observed experimentally. This discrepancy takes place in the vicinity of $\underline{c}^*$ where the theory is the less satisfactory and therefore it seems that our approximation is rather poor in this domain.

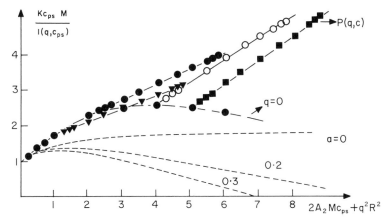

Fig. 1: Zimm plot for a mixture of homopolymers. The dotted lines corresponds to the theoretical values at $q = 0$ for increasing interaction parameter between the two polymers; $\underline{a} = (v_s/v)X_{ab}$.

On another figure, we have plotted the coefficient B of eqn (40) as function of $\underline{c}$ or $2A_2\underline{cM}$ which is a dimensionless parameter. As before, qualitatively speaking, the agreement is good but quantitatively there are discrepancies.

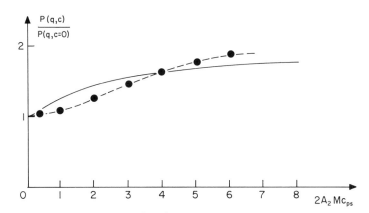

Fig. 2: The slope of the $P(\underline{q},\underline{c})$ curves as function of $\underline{c}$. The full line is the theoretical curve evaluated from eqn (40).

REFERENCES

1. H. Benoît and M. Benmouna, Polymer, 25, 1059 (1984).
2. H. Benoît and M. Benmouna, Macromolecules, 17, 535 (1984).
3. H. Benoît, W. Wu, M. Benmouna, B. Moser, B. Bauer and A. Lapp, Macromolecules, in press.
4. L.S. Ornstein and F. Zernike, Proc. Acad. Sci., 17, 793 (1914) Amsterdam.
5. H. Benoît, J. Koberstein and L. Leibler, Makromol. Chem. Suppl., 4, 85 (1981).
6. Z. Akcasu, G. Summerfield, S. Jahshan, C. Han, C. Kim and H. Yu, J. Polymer Sci. Phys. Ed., 18, 863 (1980).
7. J.S. Rowlinson, "Liquids and Liquid Mixtures", Butterworths, London (1971).
8. H. Yamakawa, "Modern Theory of Polymer Solutions", Harper and Row, New York (1971).
9. P.J. Flory, "Principles of Polymer Chemistry", Cornell University Press, Ithaca, New York (1967).
10. N.W. Ashcroft and D.C. Langreth, Phys. Rev., 156, 685 (1967).
11. W. Wu and M. Rawiso, private communications.
12. P.G. de Gennes, "Scaling Concepts in Polymer Physics", Cornell University Press, Ithaca, New York (1979).
13. W.H. Stockmayer, J. Chem. Phys., 18, 58 (1950).
14. P. Kratochvil, J. Vorlicek, D. Stakova and Z. Tuzar, J. Polym. Sci., Phys. Ed., 13, 2321 (1975).
15. A. Vrij, M.W. Van den Esker, J. Polym. Sci., Polym. Phys. Ed., 13, 727 (1975).
16. L. Leibler, Macromolecules, 15, 1283 (1982).
17. F. Joanny, L. Leibler and R. Ball, preprint.
18. T. Fukuda, M. Nagata and H. Inagaki, Macromolecules, 17, 548 (1984).

18. Polyacrylonitrile Gels in Mixtures of DMF and Water Studied by Carbon-13 NMR

Bengt Lindstedt and Per Flodin

Department of Polymer Technology, Chalmers University of Technology, Gothenburg, Sweden

Abstract – Gels formed by solutions of polyacrylonitrile (PAN) in mixtures of DMF and water have been studied by carbon-13 NMR at 30 to 95°C. Magnetic relaxation shows that gels of low water content (less than 2%) have the same segmental dynamics as solutions. Gels with high water content (12 to 14%) are turbid and show a marked decrease in signal area compared to solutions. Signal area is gained at elevated temperatures, but full intensity is not reached at the gel melting temperature. Intermediate water content (5 to about 10%) gives gels with two phases, one gel phase with segmental motions as in solutions, and one phase not visible in high resolution carbon-13 NMR.

INTRODUCTION

The thermoreversible gelation of polyacrylonitrile (PAN) in a mixture of N,N-dimethylformamide (DMF) and water has been studied in this laboratory by melting point determination (ref. 1), thermal analysis (DSC), surface area measurements and determination of exuded liquid (ref. 2, 3). The results show that at high water content (12-14%) the gels become more rigid as syneresis and gel shrinkage at drying is decreased (ref. 2). The crystalline nature of the crosslinks is confirmed by DSC, which at water contents of 6 to 10% also shows a second endotherm below the gel melting point (ref. 3). Several authors have found two endotherms in DSC or DTA (ref. 4-7). Guerrero et al could observe two different crystalline morphologies in PVC-gels by x-ray diffraction, gels that showed two endotherms in DSC (ref. 7, 8).

Carbon-13 NMR techniques can display the segmental dynamics in polymer molecules, through measurement of spin-lattice relaxation time (T_1) and spin-spin relaxation time (T_2) (ref. 9). For carbons relaxing through dipole interactions with directly bonded protons, motional narrowing is fulfilled by motions described by correlation times shorter than about 10^{-10} seconds, in the magnetic field in the NMR spectrometer used in this investigation, a Varian XL-200. In this area T_2 equals T_1 and decreases monotonously by decreasing correlation time. At correlation times longer than 10^{-9} seconds, T_1 starts to increase and T_2 continues to decrease. As T_2 is governing the intrinsic line width of a resonance line, carbons in segments having long correlation times give broader peaks in a spectrum.

Chemically crosslinked gels, swollen in good solvents, have been studied in some recent works (ref. 10-12). Results show a marked line broadening at a crosslinking density above a critical value, due to a decrease in T_2. The T_1 relaxation time is not strongly affected. The authors propose a model with a broad spectrum of correlation times (ref. 13) to describe the relaxation behaviour of the polymer networks. Physically crosslinked polysaccharide networks have also been studied (ref. 14,15) where the decrease in signal intensity from the gel is the most clear result.

EXPERIMENTAL

Polyacrylonitrile (PAN) was made in a slurry polymerization procedure described elsewhere (ref. 16). Its molar mass was determined by low angle laser light scattering in DMF to 88,700 g/mol. Reported values of water content is by volume of the solvent, and polymer content is by volume, setting the density of PAN to 1.18 g.cm³ (ref. 17). Polymer solutions were made by heating a slurry of PAN in a mixture of distilled DMF and water to 120°C until complete dissolution. The solutions were poured in 10mm o.d. NMR tubes and a 5 mm o.d. NMR tube was fitted coaxially before the gel had set. The NMR tubes were left to cool at room temperature and aged for 10 days. An external standard containing a mixture of 60 weight per cent deuterated DMSO and 40 weight per cent polyethylenoxide (PEG, MW=1500) for calibrating peak area, was added to the 5 mm tube.

NMR-spectra were collected on a Varian XL-200 operating at 50.3 MHz in pulsed Fourier transform mode for carbon-13. T_1-values were obtained with the inversion recovery method using peak maximum values (ref. 18). Relative signal areas were measured by digital integration of peaks, and usually three electronic integrations were averaged. The spectra

used here were decoupled with suppressed NOE (nuclear Overhauser effect) and the delay time
between pulses was at least 6 times the T_1 of the slowest relaxing nuclei of interest, PEG.
No mathematical line broadening was applied and usually 300 acquisitions were collected for
each spectrum. Spectra were obtained from 30 to 95°C always starting at the lowest
temperature. T_1- values and peak area were determined with ± 10 per cent and temperatures
were kept constant within ±2°C.

RESULTS

The T_1-values of gels do not differ significantly from solutions, as can be seen in Table 1.
The gel melting temperatures of a 16% gel without added water and a 12% gel with 5% water
are both 82°C (ref. 3). The shorter relaxation times in the 16% solution and gel can be
explained by a higher viscosity in that sample, and thus slightly longer correlation times.
The peaks do not show any marked broadening at the base, or at half height (Fig. 1).
Broadening of peaks is common in chemically crosslinked gels (ref. 10-12).

However, the signal area is changing with temperature. Therefore the external standard
system was developed. The signal area of the methylene peak and the sum of the three
methine peaks were calibrated to 0.371 times that of external PEG, for a 10% solution of PAN
in DMF.

The gels of low water content (less than 2%) do not show syneresis nor turbidity and
are very soft. The signal area of the gels only show a minor decrease at lower temperature
(Fig. 2). Therefore gel formation itself is not decreasing the signal area, and the
crosslinking density is not high enough to reach the critical value found by Asakura et al
in studies of chemically crosslinked gels (ref. 12). But at a moderate water addition (5 to
about 10%), the gels become stiffer, turbid and show strong syneresis (ref. 2). The signal
area is decreased at low temperatures in the 6% water gel, and in a gel with 8% water the
signal area is only 20% of maximum at 30°C (Fig. 3). The latter had to be heated to 90°C to
reach full intensity, which is above its gel melting point, 82°C (ref. 3). The increase in
signal area is spread over a wide temperature range which is mainly above the low endotherm
found in a previous DSC study (ref. 3).

At high water content (12 to 14%) the gels are cheesy and syneresis is decreased (ref.
2). These gels show a very small signal area below 50°C, less than 5% of full intensity
(Fig. 4). They do not reach full intensity at 95°C, a temperature well above the gel
melting point.

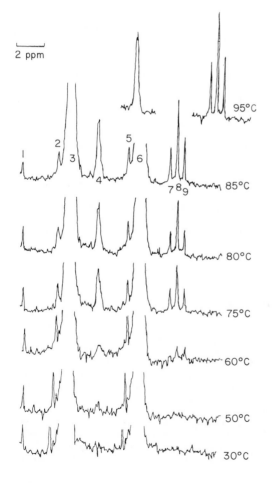

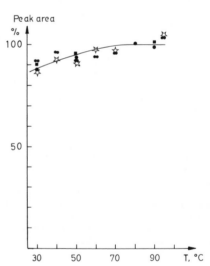

Fig. 2. Peak area vs temperature,
low water content.
16% PAN, 0% water, methine ■ and
methylene carbon ●● .
14% PAN, 2% water, methine ☆ and
methylene carbon ● .

Fig. 1. Spectra from a gel with 5% PAN in
DMF with 10% water.
Assignments: 1) DMSO-d_6, four other peaks
not shown, 2) artifact, 3) DMF, one of the
-CH_3 carbons, 4) methylene carbon of PAN,
5) artifact, 6) DMF, the other -CH_3 carbon,
7,8,9) methine carbon of PAN, the three
triads rr, mr, mm, resp.

TABLE 1. T_1-values in seconds

Temp 1. °C	PEG	Methylene	Carbon	Methine	
			rr	mr	nm
10% PAN in pure DMF, solution					
30	0.5	0.13	0.23	0.23	0.24
80	1.5	0.27	0.47	0.50	0.51
95	3	0.35	0.57	0.58	0.59
16% PAN in pure DMF, gel					
30		0.11	0.20	0.21	0.21
50		0.14	0.25	0.26	0.28
80		0.25	0.41	0.42	0.46
95		0.32	0.60	0.56	0.60
10% PAN in DMF with 5% water, gel					
30		0.13	0.23	0.23	0.25
50		0.18	0.28	0.31	0.33
80		0.28	0.49	0.48	0.52
90		0.32	0.50	0.54	0.60

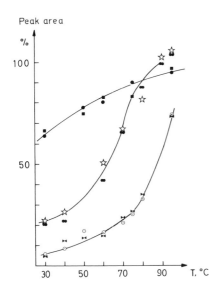

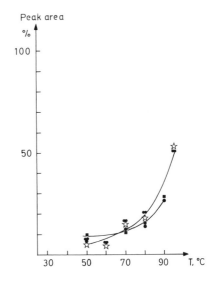

Fig. 3. Peak area vs temperature, moderate water content.
10% PAN, 6% water, methine carbons ■ , methylene carbon ● .
10% PAN, 8% water, methine carbons ☆ , methylene carbon ●● .
10% PAN, 10% water, methine carbons ►◄ , methylene carbon ○ .

Fig. 4. Peak area vs temperature, high water content.
10% PAN, 12% water, methine carbons ● , methylene carbon ■ .
10% PAN, 14% water, methine carbons ●● , methylene carbon ☆ .

DISCUSSION

We conclude that the mobility of the majority of the segments is not affected by a few short physical crosslinks. The decrease in signal area at higher water content originates from other effects than an increase in number of crosslinks, because this would decrease T_2 and increase line widths of the peaks visible in the spectra (Fig. 1). The three methine carbon peaks are well separated in solution and in gels with low water content, and are assigned to the three stereochemical triads (ref. 19, 20).

The nitrile carbon resonance line could be separated into 9 different peaks, corresponding to the 10 stereochemical pentads with overlap of the two central pentads (ref. 20). Lagerkvist and Flodin have estimated that for gels without added water the crystalline length in gel forming crystals is about 4 monomeric units and propose that those are of syndiotactic configuration (ref. 3). This length is found in two pentads, and they are found in great numbers on every chain. In gels of higher water content, there could not be observed any decrease in the signal area of those configurations. The decrease in signal area must come from the slower motion of large parts of polymer molecules, not only short segments.

PAN is found to be atactic with a probable syndiotactic diad value of 0.48 (ref. 20, 21). Triad peaks show that propagation follows Bernoullian statistics (ref. 22). Measurements on pentads do not contradict this (ref. 23). However, single crystals have been grown from solutions of PAN in propylene carbonate (ref. 24-26) and in the process of solution polymerization (ref. 27). Other authors have reported a fibrillar structure in PAN fibres (ref. 28). These single crystals or fibrils must also include material with atactic structure, as the PAN samples used are mainly atactic.

The loss of signal area in gels with more than 8% water is explained by the formation of a phase with much lower mobility than in solution, and hence longer correlation times resulting in very broad peaks not detectable in high resolution carbon-13 NMR. When the gels are heated the mobile phase grows, and thus the immobile phase is "melting". This process is independent of melting of the crystallites forming the gel network junctions, as melting occurs at high signal intensity (gel with 8% water) or at low signal intensity (about 85°C in a gel with 14% water).

The model proposed for PAN gels with high water content consists of helices of PAN aggregating and forming fibrils. The stiff fibrils stabilize the gel network and prevent collapsing of the gel structure during drying (ref.29). We propose that the immobile phase found in this study consists of those fibrils.

The transition from stiff phase to mobile phase is strongly cooperative, as the latter gives resonance lines as solutions. This indicates that there are very few connections between the two phases on every chain. At melting a great part of a PAN-chain must participate in the transition. The onset and end of these reactions cannot easily find a stereochemical explanation, as the same stereochemical configurations are found in about equal amounts in a solution of PAN and in the mobile phase of a gel of PAN in a DMF-water mixture (ref. 23).

The endotherm in DSC (differential scanning calorimeter) found by Lagerkvist and Flodin (ref.3) occurs in a lower temperature range than the fast increase in the mobile phase, as indicated by the NMR measurements. For this reason the endotherm cannot be explained by melting of the mobile phase, but is probably a transition within that phase.

REFERENCES

1. D.R. Paul, J. Appl. Polymer Sci., 11, 439 (1967).
2. P. Lagerkvist and P. Flodin, J. Polymer Sci., Polymer Lett. Ed., 19, 125 (1981).
3. P. Lagerkvist and P. Flodin, J. Polymer Sci., Polymer Phys. Ed., submitted.
    P. Lagerkvist, Ph.D. Thesis, Chalmers University of Technology, Göteborg 1981.
4. H.C. Haas, M.J. Manning and S.A. Hollander, Anal. Calorimetry, 2, 211 (1970).
5. M. Girolamo, A. Keller, K. Miyasaka and N. Overbergh, J. Polymer Sci., Polymer Phys. Ed., 14, 39 (1976).
6. H. Bergmans, F. Govaerts and N. Overbergh, J. Polymer Sci., Polymer Phys. Ed., 17, 1251 (1979).
7. S.J. Guerrero and A. Keller, J. Macromol. Sci.-Phys. 20, B, 167 (1981).
8. S.J. Guerrero, A. Keller, P.L. Soni and P.H. Geil, J. Polymer Sci., Polymer Phys. Ed., 18, 1533 (1980).
9. J.R. Lyerla, Jr, and G.C. Levy, in Topics in Carbon-13 NMR Spectroscopy, Vol 1, p. 81 (G.C. Levy, Ed.), Wiley-Interscience, New York (1974).
10. K. Yokota, A. Abe, S. Hosaka, I. Sakai and H. Saito, Macromolecules, 11, 95 (1978).
11. W.T. Ford and T. Balakrishnan, Macromolecules, 14, 284 (1981).
12. T. Asakura, K. Susuki and K. Horie, Macromol. Chem., 182, 2289 (1981).
13. J. Schaefer, in Ref. 9, p. 175.
14. H. Saito, E. Miyata and T. Sasaki, Macromolecules, 11, 1244 (1978).
15. F.M. Nicolaisen, I. Meyland and K. Schaumburg, Acta Chem. Scand., 34, B, 579 (1980).
16. Macromolecular Synthesis Coll., Vol. 1, p. 167 (J.A. Moore Ed.), Wiley, New York (1977).
17. W. Fester, In Polymer Handbook, 2nd Ed., p. V-37 (J. Brandrup and E.H. Immergut Ed.), Wiley, New York (1975).
18. Ref. 9, p. 100.
19. J. Schaefer, in Ref. 9, p. 151.
20. H. Balard, H. Fritz and J. Meybeck, Macromol. Chem., 178, 2393 (1977).
21. K. Matsuzaki, M. Okeda and T. Uryu, J. Polymer Sci. A-1, 9, 1701 (1970).
22. J.C. Randall, Polymer Sequence Determination. Carbon-13 NMR Method, Academic Press, New York (1977).
23. B. Lindstedt, unpublished results.
24. V.F. Hollander, S.B. Mitchell, W.L. Hunter and P.H. Lindenmeyer, J. Polymer Sci., 62, 145 (1962).
25. J.J. Klement and P.H. Geil, J. Polymer Sci. A-2, 6, 1381 (1968).

26. G.N. Patel and R.D. Patel, J. Polymer Sci. A-2, **8,** 47 (1970).
27. F. Kurmann, T. Kajiyama and T. Takayanagi, J. Crystal Gr., **48,** 202 (1980).
28. S.B. Warner, D.R. Uhlmann and L.H. Peebles, J. Mater. Sci., **14,** 1893 (1979).
29. S.J. Guerrero, A. Keller, P.L. Soni and P.H. Geil, J. Macromol. Sci.-Phys., **20,** B, 161 (1981).

19. High Performance Electrophoresis – An Alternative and Complement to High Performance Liquid Chromatography

Stellan Hjertén and Ming-de Zhu

Institute of Biochemistry, Biomedical Center, Uppsala University, P.O. Box 576, S-751 23 Uppsala, Sweden

Abstract – High-performance electrophoresis is performed in narrow glass capillaries (inner diam: 0.05 - 0.3 mm; wall thickness: 0.1 mm) for efficient dissipation of Joule heat in order to suppress thermal deformation of the solute zones at the high field strengths used (100 - 300 V/cm). The zones are detected by an on- or off-tube (UV) monitor. In the latter case the solutes are transferred to a conventional HPLC detector by a hydrodynamic buffer flow as they migrate electrophoretically from the capillary tube. This migration-elution technique permits collection of the purified solutes for further studies.
High-performance electrophoresis (HPE) mimics high-performance liquid chromatography (HPLC) in the sense that it permits rapid, high-resolution separations of low- and high-molecular weight substances on both an analytical and a preparative scale.

INTRODUCTION

During the last ten years a steadily increasing number of laboratories have invested in equipment for high performance liquid chromatography (HPLC) in spite of the relatively high costs. The main reason for the great popularity of this technique is that it gives higher resolution and permits higher flow rates (= shorter run times) than conventional low-pressure chromatography. We have previously pointed out that electrophoresis performed under certain conditions has the same attractive characteristic features as HPLC, namely, high resolution and short analysis times, provided that time-consuming staining or derivatization procedures for detection and quantification are avoided (ref. 1-3). This version of electrophoresis was termed "high-performance electrophoresis" (HPE).

Optimal Conditions for High Performance Electrophoresis

The experimental conditions under which the experiments are conducted in order to achieve high resolution, fast separations and rapid detection are briefly as follows (ref. 1).
The electrophoresis tube should have a small inner diameter (0.05-0.3 mm), and a thin wall (around 0.1 mm) for rapid dissipation of the Joule heat in order to suppress thermal deformation of the electrophoretic zones. For the same reason, the electrophoresis tube should be of a material of high thermal conductivity. From this point of view glass tubes are better than plastic tubes. Plastic tubes also lack the necessary mechanical rigidity at the small wall thickness required (0.1 mm) and have the further disadvantage that hydrophilic gels (such as agarose or polyacrylamide) adhere poorly to them. If an electrophoresis experiment is performed under the above conditions, high field strengths (about 300 V/cm), permitting fast runs, can be used without observable or significant thermal distortion of the zones, particularly if the tube is actively cooled. In order to record rapidly the zones all staining procedures should be avoided. We have therefore used one of the following two approaches.

On-Tube Detection

As the zones migrate along the electrophoretic tube they pass a stationary UV beam directed onto a photomultiplier connected to a recorder. UV-absorbing zones are thus recorded as peaks on the recorder chart (ref. 1, 2). Similar UV-detection techniques have previously been described (ref. 4-6).

Detection by a Migration-Elution Technique

The solutes (the zones Z) migrate electrophoretically down the capillary glass tube T (often containing a gel) and into an elution chamber C (Fig. 1). A continuous buffer flow, created

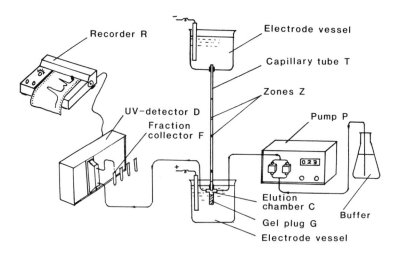

Fig. 1.   A schematical diagram of the migration-elution detection arrangement.
For details, see the text.

with the aid of an HPLC pump P, transfers the solutes from the elution chamber C to a UV-
detector D (or any other kind of detector) connected to a recorder R and further to a
fraction collector F. This detection technique - briefly outlined in ref. 3 - differs from
that previously used for preparative electrophoresis (ref. 7-14) in that the volume of the
flow cuvette is of the same order of magnitude as that of the electrophoresis tube (1-10 µl)
(and not just a small fraction of it).

EXPERIMENT AND RESULTS

HPLC has been used successfully for the separation of low-molecular weight compounds, and in
recent years also of macromolecules. We will show below that HPE is also applicable
with both categories of substances.
    A sample consisting of ephedrine (peak 1; 3µg/µl), protriptyline (peak 2;3µg/µl),
phenylpropanolamine (peak 3; 3 µg /µl), codeine (peak 4; 0.1 µ g/µl), benzylamine (peak 5;
0.05 µ g/µl) was electrophoresed into a polyacrylamide gel of the composition $\underline{T}$=6%; $\underline{C}$=3% cast
in a glass tube of length 150 mm and inner diameter 0.3 mm (for the definition of the
parameters $\underline{T}$ and $\underline{C}$, see ref. 15).   A 0.1 M Tris-HAc solution (pH 8.6) was used as buffer.
The experiment was conducted at 0.7 mA (1000 V) without active cooling of the electro-
phoresis tube.   The sample components were recorded by the "on-tube method" at a wavelength
of 250 nm after they had migrated a distance of 80 mm.   The electropherogram is shown in
Fig. 2.   A nearly baseline separation was achieved within 20 minutes.

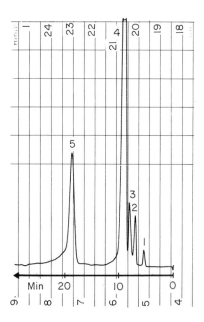

Fig. 2. High performance electrophoresis of an
artificial mixture of benzylamine (peak 5) and
some drugs (peaks 1-4).   The zones were detected
as they migrated electrophoretically past a
stationary UV detector (on-tube detection).

An example of an HPE run of biopolymers is presented in Fig. 3. The sample consisted of a commercial ovalbumin preparation, the homogeneity of which could be questioned from HPLC experiments. The water-cooled electrophoresis glass tube had the dimensions 0.25 mm (i.d.) x 200 mm, and contained a polycrylamide gel ($\underline{T}$=3%; $\underline{C}$=4%). The run was conducted in a 0.01 M sodium phosphate buffer, pH 8.0. A field strength of 175 V/cm gave a current of 0.16 mA. The electropherogram in Fig. 3 is based on the migration-elution detection technique with UV-measurements at 220 nm. It is evident that the ovalbumin preparation was not homogeneous. The analysis time was around 15 minutes.

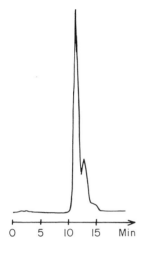

Fig. 3. Analysis by high performance electrophoresis of the purity of a commercial ovalbumin preparation. The zones were detected by the technique outlined in Fig. 1 (the migration-elution technique).

DISCUSSION

The migration elution technique described in Fig. 1 is very versatile in the sense that it permits the use of any detector employed in HPLC, for instance those based on UV-absorption, fluorescence and refraction and the spectrum of each zone can be measured on line by means of an HPLC photodiode array spectrophotometric detector.

An advantage of the migration-elution detection technique in comparison with the on-column detection method is that it permits recovery of the solutes. High performance electrophoresis therefore now parallels HPLC also in the sense that it can be employed not only for analytical but also for preparative runs on a microscale.

Although this paper is not a final report, but rather a report on work in progress, the results presented herein (see Figs. 2 and 3) and those in refs. (1-3) clearly indicate that high performance electrophoresis (HPE) has the same characteristic features as high performance liquid chromatography (HPLC); namely, high resolution, short run times, recovery of the solutes if desired and no delay in recording of the chromatogram/electropherogram after completion of the run. HPE should therefore be an attractive complement or alternative to HPLC.

A comparison of the two detection techniques

The reason why we employ thin-walled electrophoresis tubes (around 0.1 mm) is not only because they permit a rapid transport of the Joule heat but also because they have high UV transmission down to about 250 nm, even if they are made of ordinary glass. Accordingly, one need not use the brittle and more expensive quartz tubes for the on-tube detection method. Since the inner diameter of the electrophoresis tubes is only 0.05-0.3 mm it is difficult to make the UV beam sufficiently narrow to pass only the central part of the tube. The relationship between the absorbency and concentration of the solute is therefore relatively complex. Under certain conditions there is, however, an approximately linear relationship between these parameters (see ref. 4, pp. 182-190).

In the elution-migration technique described above the flow cell volume is larger than the volume of a zone (Z in Fig. 1) in the electrophoresis tube. For zones that are so well separated in the electrophoresis tube that the distance between them exceeds their widths, one can choose a pump speed (dilution factor) such that each zone can at some point be contained entirely within the cuvette without overlap with other zones. When this condition is fulfilled the maximal recorder deflection will be proportional to the total amount of material in the zone and area measurements will therefore be unnecessary.

However, for zones that are closer together, the pump speed (dilution factor) must be correspondingly increased so that two adjacent zones are not present in the cell at the same time. Under these conditions the volumes of the diluted zones may be greater than the volume of the cell and peak areas must be measured for accurate quantification. A disadvantage of the elution-migration technique in preparative (but not analytical) runs is that the zones will be appreciably diluted. However, the scale of the system is so small that the final zone volumes are still very low. In principle this problem can be solved by making the cuvette smaller, but this requires a detector of higher sensitivity and stability than those now commercially available (the detection problems are similar to those met in microbore HPLC.

ACKNOWLEDGEMENTS

The authors are much indebted to Dr. David Eaker for useful discussions and to Mrs. Karin Elenbring who skilfully performed the experiment shown in Fig. 2 and Mr. Göte Eriksson, Mr. Per-Axel Lidström, and Mr. Hans Pettersson for manufacturing the equipment for high performance electrophoresis. The drugs used in the experiment shown in Fig. 2 were kindly supplied by Dr. K. -G. Wahlund, Institute of Analytical and Pharmaceutical Chemistry. The work has been financially supported by the Swedish Natural Science Research Council and the Wallenberg Foundation.

REFERENCES

1. S. Hjertén, J. Chromatogr. **270**, 1-6 (1983).
2. S. Hjertén, in Electrophoresis ´83 (H. Hirai, Ed.), pp. 71-79, Walter de Gruyter, Berlin (1984).
3. Ming-de Zhu and S. Hjertén, in Electrophoresis ´84 (V. Neuhoff, Ed.), Proc. Fourth Meeting Internat. Electrophoresis Soc. in Göttingen (1984), pp. 110-113. Verlag Chemie, D-6984 Weinheim (1984).
4. S. Hjertén, Chromatogr. Rev. **9**, 122-219 (1967).
5. L. Arlinger and R. Routs, Sci Tools **17**, 21-23 (1970).
6. J.W. Jorgensen and K. DeArman Lukacs, J. Chromatogr. **218**, 209-216 (1981).
7. J. Porath, E.B. Lindner and S. Jerstedt, Nature **182**, 744-745 (1958).
8. L. Vargas, K.W. Taylor and P.J. Randle, Biochem. J. **77**, 43-46 (1960).
9. U.J. Lewis and M.O. Clark, Anal. Biochem. **6**, 303-315 (1963).
10. S. Hjertén, J. Chromatogr. **11**, 66-70 (1963).
11. D. Racusen and N. Calvanico, Anal. Biochem. **7**, 62-66 (1964).
12. T. Jovin, A. Chrambach and M.A. Naughton, Anal. Biochem. **9**, 351-369 (1964).
13. S. Hjertén, S. Jerstedt and A. Tiselius, Anal. Biochem. **11**, 211-218 (1965).
14. S. Hjertén, S. Jerstedt and A. Tiselius, Anal. Biochem. **27**, 108-129 (1969).
15. S. Hjertén, Arch. Biochem. Biophys., Suppl. **1**, 147-151 (1962).

20. Group Specific Protein Adsorbents Based on "Thiophilic" Interaction

J. Porath, M. Belew, F. Maisano and B. Olin

Institute of Biochemistry, Biomedical Center, Uppsala University, Box 576, S-751 23 Uppsala, Sweden

Abstract - Two new kinds of affinity gels have been prepared. They exert surprising, hitherto unknown interactions with proteins which may be used for extremely efficient group fractionations, demonstrated here with human serum as a model mixture. The protein-adsorbent interaction is tentatively called "thiophilic" referring to the presence of a sulphonic and a thioether group in the adsorption centre (ligand). A characteristic feature of thiophilic adsorption is its dependence on both the high concentrations and the constituent cation of water-structure forming salts such as potassium and magnesium sulphate. The results illustrate an inherent difference between "thiophilic" and "hydrophobic" adsorbents. Thus, the thiophilic gels may be used for selective salting-out chromatography as demonstrated with a tandem bed arrangement.

The secondary and higher orders of protein structure are stabilized predominantly by charge-charge attraction, hydrophobic interaction and hydrogen bonding. In addition, coordinative metal bonding and other minor or less known interactions can contribute further to be structure-stabilizing forces in some proteins. The spatial distribution, accessibility and abundance of the various groups that contribute to the structural stabilization of proteins have been, and continue to be, exploited in the design of newer and effective separation methods to augment or extend the already existing ones.

The introduction of the first successful and widely acclaimed procedure for immobilization of ligands on solid supports (ref. 1) gave a strong impetus not only to the development of affinity chromatography methods in general but also to the exploration of new concepts which led to the development of several novel adsorption methods. Certain of these, e.g. hydrophobic (ref. 2-5) and charge-transfer (ref. 6) chromatography were developed deliberately to exploit the subtle differences that exist between various proteins with respect to their structure-stabilizing forces. Progress in the development of methods for ligand immobilization and exploratory chromatographic experiments with the resulting adsorbents have lead in particular instances, to the discovery of hitherto unknown interactions between proteins and ligands. The bases for such interactions are sometimes not clearly understood. A case in point is a new adsorption phenomenon which we recently discovered. Its main features will be presented briefly in this communication.

The new adsorption method is based on the interaction of proteins with immobilized ligands that invariably contain a thio-ether-sulphur and an adjacent sulphone group as integral constituents. We have used the term "thiophilic" interaction to distinguish this method from related ones and to emphasize the crucial contributions of the specific groups in the ligand to the observed phenomenon. The following general structure is common to all of the "thiophilic" adsorbents we have studied so far:

$$SO_2-(CH_2)_{\underline{n}}-S-R \tag{1}$$

Variations in the structure of R were profoundly to influence the capacity and/or selectivity of the adsorbent without significantly altering its main "thiophilic" character. This is illustrated by the following series of "thiophilic" adsorbents which we have synthesized and used in some orienting adsorption experiments:

$$(P)\ -O-CH_2-CH_2-SO_2-CH_2-CH_2-S-CH_2-CH_2OH \tag{2}$$

"T-gel"

$$(P)\ -O-CH_2-CH_2-SO_2-CH_2-CH_2-S \overset{N-N}{\underset{\underset{H}{N}}{\parallel\ \parallel}} \tag{3}$$

"Tr-gel" (triazole-S-gel)

$$\text{P} \quad -O-CH_2-CH_2-SO_2-CH_2-CH_2-S- \bigcirc_N \qquad (4)$$

"P-gel" (pyridine-S-gel)

In this report, we have used Sepharose 6B (Pharmacia Fine Chemicals AB, Uppsala, Sweden) as the polymer matrix P. Other hydroxylic polymers can also be used, provided they can withstand exposure to extreme alkaline conditions.

The synthesis of these derivatives is straightforward and is achieved simply by coupling the desired thiol compound to a suitable polymer that has been activated with divinylsulphone (ref. 7) in alkaline medium. The details of the synthetic procedure will be published elsewhere.

Adsorption to the "thiophilic" adsorbents is markedly enhanced by the presence of high concentrations of one or other of a series of inorganic salts. This "salting out" effect (refs. 4, 8) is apparently a necessary requirement for adsorption to take place, a situation analogous to that in hydrophobic interaction chromatography (refs. 4, 5) on uncharged hydrophobic adsorbents. Depending on the nature of the salt, the qualitative and/or quantitative adsorption profiles obtained can be significantly altered. Consequently, one has several options to choose from in order to optimize the group fractionation of a given sample simply by choosing the "right kind" of salt. This will be illustrated by the results obtained in the following series of experiments. The results also illustrate the inherent differences between "thiophilic" and hydrophobic adsorbents, although adsorption or desorption of solutes is performed in almost identical fashion.

A series of three independent experiments were performed (see Figs. 1-3) in order to:

I. compare the adsorption properties of the three types of thiophilic adsorbents,
II. investigate differences or similarities between these adsorbents and a hydrophobic adsorbent, and
III. Investigate the influence of various salts on the qualitative and/or quantitative adsorption of proteins.

Four columns (1 x 20 cm) were used in each series of experiments and were connected in tandem as shown in Figs. 1-3. Three of the columns were packed with each of the "thiophilic" adsorbents schematically shown in the diagram above (see structures 2-4). The fourth one was packed with a hydrophobic adsorbent (H-gel), an improved octyl-agarose derivative the synthesis of which will be published elsewhere. Each column had a total bed volume of 10 ml.

The tandem column was equilibrated with 0.1 M Tris-HCl, pH 7.6, containing either 4 M NaCl, 0.5 M K_2SO_4 or 1 M $MgSO_4$. The sample of serum was dialysed against the appropriate equilibrating buffer prior to application. In each case, 6-10 ml serum (total A_{280} about 645 units) was applied to the first column in the series (Tr or T gel) at a flow rate of 15 ml/h. The effluent was monitored continuously with an LKB:Unicord (LKB-Produkter, Bromma, Sweden) and fractions of about 3 ml were collected. After elution of all unadsorbed material with the equilibrating buffer, the columns were disconnected and each was eluted with:

I. equilibrating buffer without added salt (desorption buffer I) and,
II. 30% iso-propanol dissolved in the above buffer (desorption buffer II).

The appropriate fractions were pooled and read at A_{280} nm to determine the distribution of adsorbed and unadsorbed material relative to the total number of adsorbance units applied. The results are shown in Figs. 1A-3A. Each pooled fraction was further analysed by gradient gel eletrophoresis on PAA 4/30 gels (Pharmacia Fine Chemicals) to assess the qualitative differences or similarities between the patterns of the adsorbed proteins on each of the columns. The results are shown in Figs. 1B-3B.

The results in Figs. 1A and 2A shows that; i) the amount of material adsorbed by the tandem column in the presence of 4 M NaCl (about 57%) is significantly higher than in the buffer containing 0.5 M K_2SO_4 (45%); ii) the H-gel adsorbs more proteins (30%) than the total adsorbed by the three "thiophilic" gels together (28%) in the presence of 4 M NaCl. The adsorption capacity of the H-gel is, however, drastically reduced in the presence of 0.5 M K_2SO_4 (Fig. 2A), whereas the adsorption by the Tr-gel is markedly increased under these conditions (the Tr-gel adsorbs in fact about 83% of the total material that is adsorbed by the three "thiophilic" gels together); iii) the P-gel, like the H-gel, adsorbs much more protein in 4 M NaCl (14%) than it does in 0.5 M K_2SO_4 (4%), whereas the adsorption by the T-gel is not markedly affected by variations in the type of salt used here.

These results indicate that the capacity of the three "thiophilic" adsorbents is comparatively higher in K_2SO_4 than in NaCl, whereas the reverse is true for the H-gel. They also indicate that, of the three "thiophilic" adsorbents, the T- and Tr-gels have very similar adsorption characteristics.

These conclusions are supported further by the electrophoretic patterns obtained for the adsorbed fractions from each of the four columns (Figs. 1B and 2B). The patterns show that the T- and Tr-gels have overlapping and distinct affinities toward serum proteins, whereas the P-gel has a wider specificity and even adsorbs some albumin in the presence of 0.5 M K_2SO_4. The P-gel thus behaves as a mixed thiophilic-mild hydrophobic adsorbent. Despite these differences, however, immunoglobulins and α_2-macroglobulins are adsorbed (confirmed by immuno-electrophoretic analyses) by all three "thiophilic" adsorbents. The H-gel appears to be an efficient adsorbent for albumin.

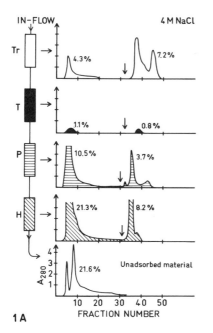

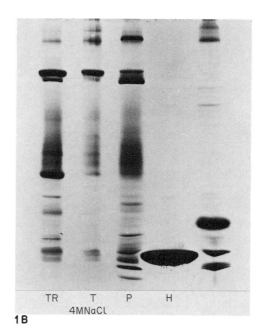

1A

1B

Fig. 1A. Tandem arrangement of the four columns connected in series. The Tr, T and P columns represent the thiophilic adsorbents and the H column represents the hydrophobic adsorbent. The curves at the right show the elution profiles of the material desorbed from each column by desorption buffers I (first peak) and II (second peak). For details, see text. The equilibrating (adsorption) buffer used here is 0.1 M Tris-HCl, 4 M in NaCl, pH 7.6. The curve at the bottom shows the material that is eluted unadsorbed from the tandem column. The percent distribution, relative to the total applied material, is indicated in each case.

Fig. 1B. Gradient gel electrophoretic patterns of the material eluted by desorption buffer I (the first peak in each chromatogram, see Fig. 1A). The electrophoretogram of the unadsorbed material is shown in lane 5.

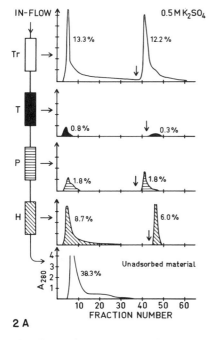

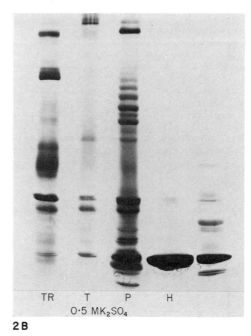

2A

2B

Fig. 2A. The experimental details and sequential arrangement of the adsorbents are identical to those shown in Fig. 1A except that the 4 M NaCl in the equilibrating buffer is replaced by 0.5 M K_2SO_4.

Fig. 2B. Gradient gel electrophoretic patterns of the fractions corresponding to those shown in Fig. 1B.

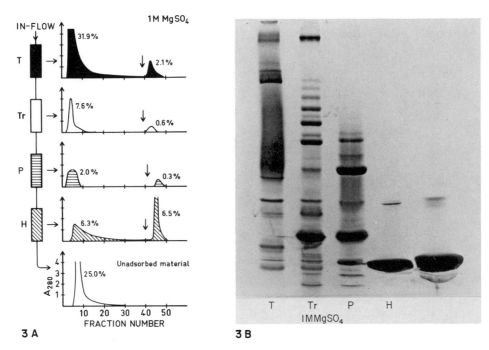

3 A   **3 B**

Fig. 3A. The experimental set-up here is identical to the previous ones (see Figs. 1A and 2A) except that; 1) the T column is placed first in the series and 2) the antichaotropic salt added to the equilibrating buffer is 1 M $MgSO_4$.

Fig. 3B. Gradient gel electrophoretic patterns of the fractions corresponding to those shown in Figs. 1B and 2B.

The results in Fig. 3A show that the T-gel now adsorbs about 77% of the total material that adsorbs to the three "thiophilic" gels together. This is consistent with the hypothesis that the T- and Tr-gels have similar adsorption characteristics. The electrophoretic patterns (Fig. 3B) further show the high selectivity of the T-gel compared to the Tr- and P-gel. The specificity of the H-gel is not affected. We would also like to point out that the most efficient group separation is obtained when $MgSO_4$ is included in the equilibrating buffer.

In conclusion, it is relevant to emphasize that the basis for the distinctive "thiophilic" interaction is not yet clearly understood. The sites on the adsorbed proteins that interact with the immobilized ligand also remain to be elucidated. In spite of this, however, the results obtained here strongly suggest that the selectivity of the "thiophilic" adsorbents most probably resides in the thioether-sulphone structure which is common to all the three adsorbents used here. The heteroaromatic ring system, including the exocyclic sulphur, modifies the "thiophilic" adsorbents to such an extent that their selectivity decreases. It is also worthwhile to stress that the specificity of the "thiophilic" adsorbents for certain classes of serum proteins is remarkable and is in many respects distinct from the various group adsorbents described heretofore.

ACKNOWLEDGEMENTS

This work has been financially supported by grants from the Swedish Natural Science Research Council, The Swedish Board for Technical Development, Erna and Victor Hasselblad Foundation and LKB-Products AB.

REFERENCES

1. R. Axén, J. Porath and S. Ernback, Nature **214**, 1302 (1967).
2. R.J. Yon, Biochem. J. **126**, 765 (1972).
3. Z. Er-el, Y. Saidenzaig and S. Shaltiel, Biochem. Biohys. Res. Commun. **49**, 383 (1972).
4. J. Porath, L. Sundberg, N. Fornstedt and I. Olsson, Nature **245**, 465 (1973).
5. S. Hjertén, J. Rosengren and S. Påhlman, J. Chromatogr. **101**, 281 (1974).
6. J. Porath and K. Dahlgren Caldwell, J. Chromatogr. **133**, 180 (1977).
7. J. Porath and R. Axén in Methods in Enzymology (K. Mosbach, Ed.) Vol. **44**, p. 19 (1976).
8. A. Tiselius, Arkiv. Kem. Geol. **26B**, No. 1 (1948).

VI The Svedberg — A Man of many Interests and Talents

21. Svedberg as an Author

Nils Gralén

Chalmers University of Technology, Gothenburg, Sweden

The scientific writings of The Svedberg were in several languages. The first were in German (1905), which was the prime language for chemical science in Europe at that time. Soon after he also published in Swedish, but only secondarily, and his papers were often produced in parallel in both languages. His doctoral thesis, in December 1907, was in German. His linguistic interest is shown by a special remark in the preface regarding the standardization of orthography and terminology in accordance with authoritative nomenclature rules by "Verein deutscher Ingenieure" and leading German chemists.

In all his scientific publications he aimed at presenting new findings that really increased scientific knowledge, and he did this in a clear, distinct way, so that facts and theories were made obvious to the reader in an intelligible way. His style was rather pedagogical, but also stimulating, because he used his imagination in comparisons and metaphors. His scientific publications number about 240, which means about 4 per year throughout his long active life in science. He was self-critical and restrained in his publications; he refrained from entering into polemics, which he considered unfruitful.

Among his scientific publications there are a few monographs. "Die Existenz der Moleküle" was published in 1912. "Colloid Chemistry", with the subtitle "Wisconsin lectures", came out in 1924. This was translated into German and also revised and enlarged by Svedberg together with Arne Tiselius in a second edition in 1928. Finally, "The Ultracentrifuge" (1940) should be mentioned, which was also published simultaneously in German. The co-author of this standard reference work, centering around Svedberg's greatest methodical achievement, was Kai O Pedersen, and several others also contributed.

Svedberg's first publication in English was in 1909, with the title "The Brownian Movement". In it, he mentions his gratitude to a translator, Dr. Elkan Weschler. Later on, English became his most frequently used language for publication, especially after World War I, and it became almost a second mother tongue to him. But he also used French, and from 1948, when he visited South America and Spain, he also took up Spanish, which he acquired rapidly.

Svedberg was also an inspiring critic and inspiring teacher of scientific writing to his pupils and collaborators. He always emphasized that writing should present the matter adequately, briefly and clearly. Such principles were also applied to his lectures. He gave many series of lectures to his students, covering new subjects every year. Many of us remember him lecturing in his high-pitch, almost squeaky voice, very distinctly, never using a manuscript and with no humming and hawing or difficulty in finding adequate expression. He had a splendid ability to convey his interest in his subject to his audience, and he illustrated his lectures abundantly with pictures from various publications, his own and others.

Svedberg wrote some books of popular essays in Swedish on science and related problems. Two of them, "Matter" (1912) and "The Decadence of Work" (1915), were also translated into German. They show an eminent skill in popularizing scientific discoveries and problems, from both the past and present, and his own research. His imagination and his broad knowledge made his presentation vivid, stimulating and easy to read. His writing is sometimes poetical, and during this part of his life, he also wrote in verse, although he never published these poetical attempts.

Another of his smaller, popularizing books was entitled "Research and Industry" (1918). In it, he develops for the first time his ideas about the necessity of collaboration for mutual benefit between industry and scientific research. This subsequently became the object of much endeavour on his part, especially from the 1940's. Among his publications were a great number of lectures, pamphlets, reports and appeals on this subject, always enthusiastic and inspiring. He also covered the social aspects of industrial development, e.g. in lectures on subjects like "The Aims and Means of Research", "Atomic Power - Our Most Important Energy Reserve", and "Man and the Machine". All these were given to Swedish audiences from various parts of society, among them, politicians and labour unions, school children and, via radio, the general public.

A special interest of Svedberg's was botany. But we have only a few printed reports in that field from him. In these his deep knowledge of plant species and his admiration of the beauty of nature shine through in an almost lyrical way. In a short paper entitled "Something about the Vegetation on the Sandy Fields of Eastern Skane" he writes at the end:

"The strange bright mood of the morning of creation has, it seems to me, always hovered over these sparsely flowering Elysian meads."

This brings me, finally, to Svedberg´s poetical talent. As I said before, he never published any of his poems, certainly due to his self-criticism, which must be judged overambitious by those who had the opportunity to read some of his poems (and I am one of them). Svedberg had a deep interest in and familiarity with poetry, both in Swedish and in foreign languages, especially French, and he had a great collection of it, as well as of other literature, including fiction and drama. His poetical vein is often revealed in his prose writings as well.

In one of his essays on natural science, he uses as an illustrative example what is meant by an irreversible reaction, the throwing of a stone into a lake. It is written in Swedish prose, and I will not try to present a translation but only indicate the content and its poetical presentation. Imagine the surface of a lake, smooth as glass, and reflecting the surrounding wooded countryside, and then imagine the thrown stone, the splash, the drops rising and falling back, the concentric circular waves spreading over the lake, until they reach the shores, reflecting and interfering, and finally level out until the lake is just as it was, mirroring the firs as before. And then The Svedberg describes the reverse reaction, extremely improbable but imaginable. The surface begins slowly to move irregularly, the waves collect themselves in circles, and the outer parts of the lake suddenly become smooth. The smooth surface grows inwards, as the circular waves move concentrically towards a centre, water drops release from the waves upwards and fall back, all into the same spot, the centre. When the outermost circular wave has reached the centre the whole lake is smooth again. At the same moment a stone rises from the centre and falls back into your hand, most surprisingly quite dry, as if it had never travelled to and from the bottom of a lake.

This prose poem by The Svedberg is charming and at once incites imagination, mysticism, awe of nature, and scientific knowledge. I read it many years ago and have searched for it recently. I found out that a slightly different version was used by Svedberg in a radio lecture on atoms in 1946. This section of the lecture was printed in a Swedish literary journal in a collection of poems, selected from various authors. Svedberg probably never published it himself. To me it still has the character of a beautiful dream, one that is but imperfectly recollected.

REFERENCES

N. Gralén, The Svedberg, in Swedish Men of Science, Stockholm (1952).

S. Claesson and K. O. Pedersen, The Svedberg 1884–1971, Biographical Memoirs of Fellows of The Royal Society, Vol. 18, London (1972).

S. Brohult, Theodor Svedberg. Levnadsteckningar över Kungl. Vetenskapsakademiens ledamöter, 184. Stockholm (1982).

T. Svedberg, Unpublished Writings, "Fragment", kindly placed at my disposal by Mrs. Margit Svedberg.

22. Svedberg as a Botanist

Carl-Johan Clemedson

Royal Swedish Academy of Sciences, Stockholm, Sweden

Amateur botany has a long and fine tradition in Sweden, dating back to the time of Carl Linnaeus. From the Linnaean epoch it extended to ever wider circles, and this is certainly one reason why clergymen, physicians and people of many other professions - as amateur practitioners of "the most charming of sciences", to quote Linneaus´s own words - have played such an important role in the botanical exploration of Sweden.

One of this country´s most outstanding amateur botanists was The Svedberg. I am sure that it is by no means an exaggeration to state that he was the foremost Swedish amateur botanist of his time, an ´amateur´ only in the sense that botany was not his profession.

The Svedberg entered into his lifelong botanical career at the age of seven, when in 1891 his family had moved from the manor house Flerång near the city of Gävle, where Theodor was born, to Hommelviken in Norway. There his father, who was a manager of ironworks and also had a keen interest in natural history, took him out on excursions to collect minerals, plants and insects etc. In 1893 the family moved back to Sweden again, first for a short while to Uppsala and then to the mining district Bergslagen in Västmanland, and many of the plants in Svedberg´s very comprehensive herbarium date back to the years 1894 to 1900, when he botanized not only in the vicinity of his home at the Elfvestorp and Karmansbo works but also in other parts of the province of Västmanland.

Svedberg got his senior school education at the Karolinska Grammar School in the city of Örebro, where he was lucky to have as his teacher of natural sciences a prominent botanist, Ernst Adlerz, who was an expert in the plant genera Hieracium and Rubus (blackberries) and especially in mosses. He had such confidence in Svedberg´s knowledge that he allowed the young boy to assist him in collecting mosses for a moss flora, which Adlerz published in 1907. During his school time in Örebro Svedberg also performed systematic studies of various vegetation types. His main interest was the vascular plants, but in order to botanize in the wintertime as well, he often made early morning excursions to collect stone and tree lichens before going to school. He passed the Grammar School with the highest grades in all subjects, and he also got prizes and grants, among others two advanced botanical handbooks, one German and one Danish, for his excellent knowledge of botany.

In his choice of further career and university studies he was uncertain whether his main subject should be botany or chemistry and physics, and the reason why he decided to devote his life to chemistry and physics was, as he has expressed in his autobiography, which he wrote in the years 1960 to 1961, that he had an intuitive feeling that the biological sciences needed to be refreshed by contributions from the more exact natural sciences, and that many biological problems could be explained as chemical phenomena. And so he became a world-famous physical chemist and Nobel Prize laureate but also in addition one of Sweden´s foremost amateur botanists and plant collectors.

Svedberg´s Botanical Excursions and Expeditions

During his long life Svedberg made innumerable botanical excursions and expeditions not only in Sweden and the other Nordic countries but also to many places in Europe, especially in England, Ireland, France, Spain, Switzerland, Italy and Russia and in his later years also to polar regions. In 1923 when he was a guest professor at the University of Wisconsin in Madison, he used part of his spare time for botanical studies in that state. Later on he also botanized in California and in South America. Not only on holidays but whenever he travelled to take part in scientific meetings or on business, he always took his herbarium press with him. He had quite an unusual ability to find time to make botanical field studies and to collect plants and to photograph plants and plant habitats. He kept close contact with a large number of Swedish and foreign botanists, professionals as well as amateurs, who often accompanied him on his botanical excursions.

Svedberg´s botanical interest embraced a very wide field, but I think that one is justified in maintaining that he had a special predilection for seashore flora and for subalpine and alpine plants. In this connection it should be mentioned that as early as 1922 he outlined a programme for the study of the field flora of five different districts in Norway, a programme that he was able to fulfil almost entirely. He evidently had an avid interest in the Norwegian flora, and he made a large number of excursions to many parts of that country. Late in his life he also got the opportunity to study the flora of Spitsbergen as well as of Iceland and Greenland. In August 1959, when Svedberg celebrated his 75th birthday, he got a birthday present, which as he states in his autobiography, surpassed everything that he could ever have dreamt of. A Swedish company offered him and his wife Margit a journey to Spitsbergen. The journey was realized the following summer and was botanically as well as from other points of view highly successful, not least due to the fact that they enjoyed unusually good weather, the best that Svalbard had had in 50 years. Svedberg expressed his enthusiasm over what he saw of the flora of Spitsbergen in the following words in a paper entitled "Some Impressions from a Botanical Journey to Spitsbergen": "For me who has botanized for many years in the Scandinavian mountains and after long walking-tours and hard climbing may have managed to find a few specimens of our rare snow-line flowers, it was an overwhelming experience to find the same flowers here in Spitsbergen growing by the thousands just outside our house."

The journey to Spitsbergen gave a strong new impetus to Svedberg´s great interest in arctic flora, and in 1961 he and his wife made a botanical expedition to Greenland. With excellent support from the United States Air Force and from the Arctic Station of the University of Copenhagen, they were able to do extensive botanical studies during three weeks in three different areas of the coastal belt of West Greenland, namely at Söndre Strömfjord, the Disco Island and far up north at Thule. A short account of this expedition was published by Svedberg under the title "Västgrönland - ett botaniskt paradis" (West Greenland - a Botanical Paradise). In July 1965 he and his wife returned for another botanical expedition to Greenland, a birthday present on his 80th birthday, and in connection with an expedition to Iceland in 1967 they made a third, short visit to Greenland. Svedberg´s last more distant expedition was to the Faroe Islands in 1968.

Svedberg always planned his excursions and expeditions thoroughly, and he had an exceptional ability to find the plants he was searching for. He was also very eager and daring, and some of his many excursions were not without dramatic episodes. Once in Norway, for example, he was searching for an extremely rare fern, Cystopteris sudetica, in a remote, unpopulated area at the Vinstra River. He eventually found the fern in a deep crevice and jumped down to photograph it without thinking of how he would manage to get up again. However, the brittle trunk of a fallen spruce was his rescue this time.

The Svedberg´s Herbarium

Through the years Svedberg collected quite an extensive herbarium of vascular plants. It comprises some 11,000 sheets of plants originating mainly from the Nordic countries and from Spitsbergen and Greenland. In many cases there are a number of copies of one and the same species collected at various locations in different countries. In 1966 Svedberg donated his herbarium to the Institute of Plant Biology in Uppsala, where it is kept in an excellent condition and, in accordance with his will, is called The Svedberg Herbarium.

In Linnaeus´ days a collection of pressed plants was termed a herbarium vivum. In a transferred sense Svedberg´s herbarius was to him really a herbarium vivum, a living herbarium. In order to be able to take out and work with his plants, he generally did not fix the plants to the herbarium sheets with adhesive tape or strips in the usual way but instead enclosed the plant in an envelope glued to the herbarium sheet. So, for him his herbarium was not merely a collection of dried plants but really a living working material. Svedberg not only collected and pressed plants, but he was also an extraordinarily skilful plant photographer. His botanical collections also include thousands of photographic plates and photo copies of plants and plant habitats, both in black and white, and in colour, as well as stereophotos.

Svedberg´s Botanical Publications

The Svedberg´s comprehensive list of about 250 published papers and books includes some ten botanical papers. The first two of these, both published in 1922, deal with a series of experiments in statistical vegetation analysis performed by Svedberg in 1919, in which he made a critical analysis of the statistical distribution of plants in randomly chosen square meter areas. There is a joke about the difficulties that he could encounter in such a work. He was studying the statistical distribution of the two different colours of the flowers of the plant Fritillaria meleagris, in Swedish named "kungsängslilja" after its famous habitat Kungsängen (The King´s Meadow) here outside Uppsala, where this plant was originally discovered in 1742. (Its English name is Fritillary or Snake´s Head). Well, Svedberg was walking in Kungsängen throwing the square meter frame and counting the numbers of Fritillaries with red and white flowers, respectively, and once when doing so, he eventually caught up with a lady picking Fritillaries, but only those of the one colour that she preferred.

Most of Svedberg´s botanical papers are short descriptions of new localities for rare plants discovered by him or of unusual habitats that especially interested him. Two of the papers are descriptions of his and his wife Margit´s expedition to Spitsbergen and of their first visit to Greenland. Certainly, Svedberg had much more botanical material that would have been worth publishing, but his modesty probably prevented him from doing so. This is evident from a letter to the editor of the Swedish botanical journal Svensk Botanisk Tidskrift, written when he submitted his first paper on statistical vegetation analysis for publication.

This modesty is also evident in his introduction to an enquiry paper about his most interesting botanical discoveries, in which he states that "I myself have never made any great botanical find, but I have had much pleasure in seeing nice plants".

Vretaudd

During his long life The Svedberg had a number of summer residences, where he did extensive botanizing and gardening. One of these residences was named Flava after the Latin name of one of his favourite plants, the Yellow Horned-Poppy, Glaucium flavum, which grows only rarely in Sweden, on the West Coast. He studied the occurrence of this plant in the district of Bohuslän and was especially interested in the relation between the frequency of the plant and the winter temperature, and he found a strong correlation between the January temperature and the number of flowering plants.

Another of Svedberg's vacation places was Vretaudd, situated on Lake Mälaren southwest of Uppsala. He acquired Vretaudd, a lovely area with oak-hazel groves, as early as 1933, and later on he enlarged the estate more and more. Not only did he enjoy the original wild flora there, but he also brought home and planted at Vretaudd a large number of species from other parts of Sweden as well as from foreign countries. He succeeded, for example, in introducing there the European mistletoe, Viscum album, which is difficult to seed, since its berries normally have to pass the gastrointestinal tract of birds in order to germinate. Svedberg, however, succeeded both with mistletoes from Sweden and with those he had bought in a market-place in Vienna.

Gardening was another of The Svedberg's favourite hobbies, and it is said that he felt that hardly anything could make him happier than to plant a seedling and to follow it as it grows up and develops. He also showed a keen interest in the conservation of nature, particularly of rare plants and of habitats worthy of protection, and he also succeeded in having some rare plants protected by law in some habitats in Sweden.

To conclude this short introduction to the exhibition of The Svedberg as a Botanist one is certainly justified in emphasizing that throughout his life Svedberg possessed a passionate interest in botany and an enormous knowledge of the subject, quite comparable to that of professional botanists. He had an extremely keen eye for what various kinds of habitat could offer botanically, and he succeeded in finding almost all the plants he looked for. From his youth it was his ambition to see all wild Swedish vascular plants in their natural habitats, and he was successful in doing so, but it took him almost his whole lifetime. The Svedberg was no doubt one of the greatest amateur botanists that Sweden has ever had.

THE SVEDBERG'S BOTANICAL PUBLICATIONS (all in Swedish with T. Svedberg as author)

1. Ett bidrag till de statistiska metodernas användning inom växtbiologien.
    Svensk Botanisk Tidskrift, **16**, 1–8 (1922).
2. Statistisk vegetationsanalys. Några synpunkter. Svensk Botanisk Tidskrift, **16**, 197–205
    (1922).
3. Astragalus danicus och Botrychium simplex funna i Bohuslän. Acta Horti Gothoburgensis,
    3, 167–168 (1927).
4. Mitt roligaste växtfynd (Svar på enkät). Sveriges Natur, **36**, 78–84 (1945).
5. Något om vegetationen på östra Skånes sandfält. In Natur i Skåne (B. Hanström and K.
    Curry-Lindahl, eds) p. 127–136. Bokförlaget Svensk Natur, Stockholm (1947).
6. Glaucium flavum Cr. på Stora Vassholmen i Fjällbacka skärgård. In
    Festskrift tillägnad J. Arvid Hedvall den 18 januari 1948, p. 553–562. Elanders
    Boktryckeri AB, Göteborg (1948).
7. Några intryck från en botanisk resa till Spetsbergan. ("Some Impressions from a
    Botanical Journey to Spitsbergen"). Svensk Naturvetenskap, **14**, 186–189 (1961).
8. Om Draba Gredinii E. Ekman på Svalbard. Blyttia, **19**, 158–159 (1961).
9. Västgrönland – ett botaniskt paradis. In Festskrift tillägnad Carl Kempe 80 år 1884–
    1964, p. 859–872. Almquist & Wiksell Boktryckeri AB, Uppsala (1964).
10. (with I. Svedberg), Valeriana baltica fridlyses på Lucerna. Bygd och Natur 178–181 (1939).

23. Svedberg as an Artist

Pernille Ahlström

Konstvetenskapliga Institutionen, Uppsala University, Uppsala, Sweden

The Svedberg painted. Art runs like a silken thread, sometimes taut, sometimes slack, through the tapestry of his life; at moments he pursued it with the same intensity and thirst for knowledge that characterised his entire being. He sought to glimpse Truth through the exercise of Science and to confirm the visible world about him through the knowledge of its concrete form.

Art was a release for the imagination and the need to create for The Svedberg. In his unpublished memoirs, Fragments, he wrote, quoting himself from the year 1915, that the act of creation is "the satisfaction of feeling with the world, of seeing from within the objects about one." And I would here venture to say that his comment following upon the above quote, that "the joy of becoming one with the world through the work of art" was his personal definition of the meaning of art. When Svedberg spoke of art he did not limit himself to the world of paint and canvas, but included literature as well. Although the present essay does not deal with his literary tastes or own writings, it should, at least, be pointed out that The Svedberg read extensively and amassed a large library during his lifetime. But, let us return to paint and canvas...

Through the creation of art he not only rediscovered, but gave new form to the concrete world, while, at the same time, expressing his own inner reality. In the beginning this inner reality sought its identity in Nature and was then draped in the elusive perfumes of Nietzschean thought. Later, however, with growing maturity and a more relaxed self-confidence, this identity in Nature was released from those elusive perfumes and sought fresher, more energetic scents.

The Svedberg began drawing as a small child when his family lived in Norway. The objects of his interest were animals of all kinds, including insects and plants, and even an occasional landscape. Growing up in rural Norway and Sweden, he drew the animals about him, but seems also to have been stimulated by illustrations. In many cases, explanations written in a large struggling childish hand accompany the drawings, and we are confronted with the feeling that these drawings might have a slightly deeper intent than mere doodling. The boy´s father had already in Norway begun to take him on botanical excursions and thus The´s more scientifically coloured sense of curiosity was aroused. Being an only child and thus naturally left to his own devices are factors that may have helped to stimulate this curiosity and its resultant activity. He drew in pencil and would at times fill in the contours with water-colours. His grasp of proportion commands respect and his eye for detail is delightful. Thus, already from the age of six, the future scientist was eagerly translating the world about him into the language of line and colour.

As a young man Svedberg´s ability with the brush developed and improved, and he chose to work mainly in watercolours, but with an occasional essay into the technique of egg-tempera. Oils were considered to be too time-consuming, for a work was generally a quick affair. He painted when the mood struck him, when a certain problem of composition or technique arose, or simple when Nature inspired him to work. Once, when travelling in Norway, he became so taken with the scenery that he promised himself that he would one day return and paint it; in August of 1925 he returned to Finse and the mountains of Norway to paint, producing, in his own opinion, the best he had ever done. That painting was important to him is also evident in the fact that he arranged a little studio in his apartment in Uppsala in the spring of 1916.

The landscape now replaced all other motifs, with few exceptions, and remained throughout his life the focus of his attention in painting. Once through the years of World War I, The Svedberg practised landscape in the melancholy and deeply romantic spirit prevalent in Northern turn-of-the-century art, whose roots were to be found in the symbolism of Gaugin and Munch. His landscapes during these early years are flat, often misty and lonely; their colours are soft and earthy in tone; their contours betray the influence of Art Nouveau. All of them exude a feeling of mood. They are silent in themselves, but betray the emotions of the painter in a manner that may speak of estrangement and bewilderment, yet with a certain discipline. Other scenes, equally silent, tell of a sense of serenity, perhaps granted Svedberg through the contemplation of, and participation in, Nature itself.

The year 1914 was a difficult and upsetting time for The Svedberg, and in October of that year he wrote of his own emotional state: "The world was beautiful. If he walked alone and wandered far away from the humans, then he felt it. The air, the sky, the earth, the trees; it was so infinitely beautiful, and their beauty was in their actual being... And he felt himself as a part of this reality, redeemed. And he saw beauty in the humans, but more in their works."

It is here important to remember that late 19th century Nietzschean thought was still current and that Svedberg, now a successful teacher and researcher at the Institute of Chemistry at Uppsala University, was a devoted follower of Friedrich Nietzsche. In August of 1915 the then 31-year old Svedberg wrote: "The passion to seek in the world of phenomena is a great pleasure and happiness, for, should I find anything, it will give me the illusion of being one with the world, of seeing more than others have seen... All is the desire to lose oneself, to be annihilated, to become like a grain of sand on the beach."

And then, sometime around 1918-1920, there is a break in style and the state of his emotions. His motif remains the landscape, but where he previously sought inspiration in the emotional aspects of Symbolism, he now sought it in the painting traditions of Van Gogh and Post-Impressionism, and Matisse and French Expressionism. His colours lighten and are clearer, the brush-work becomes dynamic, and the neutral ground of the paper on which the paint is applied is allowed to participate in the actual composition.

The work of Cézanne had earlier shown how the canvas itself - its colour and texture - was an active part of the artwork. Cézanne's influence can be seen in the works of Swedish artists such as Carl Wilhelmsson whom Svedberg met during the summer of 1915. And when not visiting the artists themselves in their studios or homes, he would see their works in galleries and museums, and therefore was not unaware of either past or present art. Vermeer was a great favourite of his, but the Cubism of Picasso and Braque is nowhere felt, nor is it accepted in its entirety by the avant-garde of the professional art world during this period. The truly abstract and thoroughly intellectual art of the Russian Constructivists and the Dutchman Piet Mondrian was not appreciated until the 1940's, almost thirty years after its conception. It was, instead, the influence of Cézanne and later, Matisse and French Expressionism that dominated the art scene in this country, the latter having been brought to Sweden through the works of Isaac Grünewald and his colleagues, most of them students of Matisse.

In French Expressionism (also called Fauvism) the motifs are classical; the landscape, the portrait, the still life, and scenes from the world of the café and theatre. The art of Matisse with its character of classical proportion and grace is joyous and decorative, and as late as the spring of 1948 The Svedberg renders the exoticism of Rio de Janeiro in the sweeping character and colours of that style. This classicising aspect is true, however, not only of painting, at the time of Svedberg's break in style, but characterises Swedish architecture and design in general after 1915. Nor should we forget the above-mentioned Van Gogh, whose work continued to influence Svedberg throughout his life.

Thus, that feeling of mood, of melancholy, in Svedberg's own works, of which I spoke earlier, fades and is replaced by something akin to a feeling of brightness. The tortured existentialist ponderings of the young man have softened and relaxed. Instead, that natural curiosity to grasp and document the concrete world about him is accompanied by the more mature and confident man's sense of the aesthetic. Although his landscapes remained unpopulated and showed, at most, an occasional cottage or boat, they now seemed to be more a positive reaction to the simple beauty and grandeur of Nature.

This sense of the aesthetic naturally stimulated thoughts of transferring the artistic abilities from paint to other materials. When the need for new curtains in the Institute of Chemistry at the university arose in 1954, Svedberg himself began to consider a possible design. His son, Elias Svedberg, suggested that he contact the Textile Studio at the fashionable department store NK in Stockholm. The studio was then headed by the designer Astrid Sampe, and, in cooperation with her and her colleague, the noted designer Viola Gråsten, two abstract designs were eventually developed. One represented atoms on a black background and was fittingly called "Atomics", the other, chromosomes in black on a gray ground called "Genetics". These textiles were then displayed in October of the same year at NK and Svedberg received positive reviews. But trying his hand at textile design was a brief exercise done simply for the fun of it and Svedberg barely mentions the event in his memoirs. Nor, in fact, does he ever mention that his own paintings have been publicly displayed in Uppsala at the Student Salons of 1945 and 1948.

The Student Salon of 1945 showed works not only of students but of their professors as well. A few of The Svedberg's paintings were exhibited - one in the style of Van Gogh. Three years later at the salon of 1948 he exhibited again when fresh from his trip to South America. Here it seems to have been Matisse and his school that answered the problems of form and colour. That he was much appreciated for his work is illustrated by the following commentary in the newspaper Stockholms-Tidningen on April 25, 1948: "The Svedberg, Jack-of-all-Trades, is the most interesting of the exhibiting artists... This is a versatile 64-year old with an almost frightening vitality..."

It is this versatility and "almost frightening vitality" that characterised The Svedberg in practically everything he did, that separated him from the mainstream, and drove his as a young man into the arms of Nietzsche. It is this that drove him to seek knowledge and expression of life within the sciences as well as the humanities, where he not only practised painting, but devoured the literature of his day.

Nevertheless, art was not the mainspring of his life. It was a facet, and only one of many. He did not seek his identity in it but in the pursuit of knowledge through the methods of science. It was for his discoveries within science that he received the Nobel Prize in 1926, not for any great works of art. That he himself barely bothered to mention in his memoirs his exhibitions of paintings or the reviews he received indicates his own attitude towards his painting. It was a private affair. Should the world seek to know and pass judgement on his splashes of colour on paper, then it was free to do so. As far as The Svedberg was concerned, it didn´t signify.

I would like to thank Margit Svedberg for allowing me the use of her late husband´s unpublished memoirs, Fragments. Others who have been of assistance to me in my work know who they are and I am equally grateful to them.

I would also like to point out that this essay is not absolute in character, and that the quotes taken from the Svedberg´s memoirs are my own free translations.

24. Svedberg as a Promotor of Science and Technology in Society

Nils Gralén

Chalmers University of Technology, Gothenburg, Sweden

In 1918, Svedberg published a small book in Swedish under the title "Forskning och industri" (Research and Industry). It contains a number of essays on science, several of them showing how the production industry has taken its ideas from scientific knowledge and discoveries, and thereby promoted human welfare and development in society. These essays show Svedberg's open mind regarding relations between science and industry. No doubt this has its background in The's own childhood and youth: he grew up in industrial surroundings; his father was the manager of an iron works and a technical consultant.

But it was not until 1933 that the indication came that Svedberg's methods of scientific research could be of use for industry. This happened when he was invited to lecture at the club of technologists in Falun, where Sweden's (and the world's!) oldest industrial company, Stora Kopparbergs Bergslag, was the host. He lectured under the title "From the research fields of the ultracentrifuge", and concentrated the presentation on proteins, which were then his main interest, but he also mentioned the possibility of using the ultracentrifuge for his main interest, but he also mentioned the possibility of using the ultracentrifuge for studies of cellulose and other high-molecular organic substances. This was taken up in the discussion after the lecture, and the company president, Mr. Lundquist, disclosed an interest on the part of industry to know more about cellulose and why it is able to form paper. Svedberg responded positively, but the cost of research prevented closer collaboration at the time. A few years later, in 1936, several cellulose industries in Sweden collectively started support of cellulose research at Svedberg's institute of physical chemistry. At about the same time the brewing industry supported research on beer and barley proteins. Svedberg was very active in shaping the contacts and organizing this research - in a rather informal way, because such collaboration was quite new for universities at that time, and no centrally devised rules defined its boundaries or possibilities. The aim was to increase knowledge about the materials and products of industry - not to develop new materials or products or to increase efficiency, in other than indirect ways: when the technologists in industry learnt more about what they were working with, this would give them better opportunities to develop better products and processes. During World War II Svedberg engaged his institute more directly in industrial support by developing a method for production of synthetic rubber, and even building a pilot plant for that purpose in a corridor of the institute. Sweden suffered from lack of materials due to the trade blockade, and Svedberg was also engaged in problems of development of other materials, such as nitrocellulose and viscose rayon.

Svedberg's combined interest in research and industry soon became more widely known and appreciated in industrial circles, and in 1942 three industries in Stockholm found it profitable to create a research laboratory of their own, called the LKB-laboratory after the initials of the companies. Svedberg's collaborator Sven Brohult was made director of this laboratory. It is still an active organization. In the following year a company was formed, LKB Produkter AB, for the production and marketing of scientific instruments. Svedberg's main interest in this company was the production of instruments developed at his own institute, but it was later found that such a basis was not sufficient. This company, which is our host today, now has a wider interest, to supply instruments and other utilities to a wide range of customers, namely for medical and biochemical research, and not only to produce instruments for a special research field.

After a petition from the Royal Academy of Engineering Sciences and the Swedish Association of Industries in 1938, the Minister of Commerce appointed a committee to investigate measures to be taken for the promotion of technological research. The Svedberg became a member, and his ideas played a leading role. This committee became one of the most fruitful governmental committees in Sweden. After its chairman it was called the Malm committee and it resulted among other things in the first research council of Sweden, the Council for Technical Research, which started its work in 1942. Svedberg was one of its members and had a dominating role in its work, up till 1957, when he resigned at 73 years of age. The council had a budget of 700 000 Sw.kr. during its first year. Its present successor, the Swedish Board for Technical Development, has a budget of about 600 million Sw.kr.

The Malm committee also proposed and achieved the creation of several branch institutes, for research into technology of special branches of science. Svedberg was especially active in two of these, namely for food preservation and for textiles. He had the intention of placing two of his pupils as heads of these institutes, but one of them chose to become a teacher instead. The second, the textile research director, is now addressing you and can verify how great an interest Svedberg took both in the creation of the textile research institute and its research, and as a member of its board for several years.

A talk about Svedberg as a collaborator with industry would be incomplete if it did not mention Gustaf Werner and the institute that was named after him. Through Professor John Naeslund, Svedberg got in contact with the rich bachelor, donator, land speculator and textile merchant in Göteborg, Gustaf Werner, and convinced him to give nearly 2 million Sw.kr. for the design and construction of a cyclotron for research. Werner had two interests therein. One was the possibility of catalysing the synthesis of big molecules (textile fibres) by radiation, the other was the medical use of strong radiation. Svedberg already had a great interest in radioactivity from his youth, during the time of the Curies' research activities, and now, at more than 60 years old, his research efforts turned to radiation and to investigating what could be performed with the cyclotron. And he was successful in this field as well. Beside his own research he also acted as a member of the Atomic Research Council, and for the promotion of nuclear power, which he considered essential for our energy supply.

Svedberg's promotion of science and technology in society was always based on his belief in science and its necessity to humanity. Technology is one of the fruits of science, for human development and social welfare. But technology and industry are not the only goals for science, not even for the natural sciences. Science is an essential factor in society in itself; it is a basic ingredient of culture, and therefore it is in the interest of any country's government to give fair resources and possibilities to research activities. Each nugget of knowledge has a positive value. But Svedberg was also aware of the risks of misuse of scientific resources and discoveries, for instance for war purposes. We have no other guarantee against misuse than further education for responsibility. He also stressed the necessity of investigating man as a social being: research into psychology, sociology and political sciences. His views are expressed in a lecture from 1947: "The Aims and Means of Research". There he says on humanistic and social sciences: "Intensive activity for enlightenment in these areas must sooner or later bear fruit. If we become clearly aware of the nature and causes of our affects, we ought to be able to control them more easily. But, as we all know, it is no simple task we have before us. Every nation as well as every individual must learn to curb his urge to demand for himself without consideration for others, must be freed from the constant fear of losing what he possesses. Such a change of heart can only be achieved through deep insight into the causes of our spiritual complexes, through understanding natural phenomena, within as well as outside ourselves, and through insight about the greater whole. We are lost if we cannot exploit the tools of research for the common good and learn to strive toward common goals."

Through his scientific work, but also through his promotion of science and technology in society, Svedberg marks an epoch.

REFERENCES

T. Svedberg, The Aims and Means of Research, IVA **18**, 127 (1947). Translation of address delivered at the official opening of the laboratory of Höganäs-Billesholms AB on Sept. 10th, 1947). A new translation of this lecture is published as Chapter 25 of this volume.

VII The Svedberg's Views on Science, Technology and Society

25. The Aims and the Means of Research

The Svedberg

Institute of Physical Chemistry, Uppsala University, Uppsala, Sweden*

What one usually means by research, I suppose, is the systematic questing for enlightenment. Our entire material culture now rests on such questing. Unfortunately the results of research can be exploited in the service of both peaceful progress and of war. The sounding of the secret depths of the human soul have unfortunately yielded few results: today we are as powerless as we were a thousand years ago to master the powers of destruction. It is small consolation that the search for tools of destruction and defense have often yielded as by-products exceedingly valuable peaceful aids: as examples of this I can mention the development of flight after the First World War, and the utilization of atomic energy and micro-waves during the recent World War (in atom bombs and radar). Our mastery of these latter forms of energy have not as yet noticeable influenced our daily lives, but, in time, the transformation of our existence which the domestication of atomic energy in particular must entail will be tremendous. For experimental physics (and thereby for the theoretical branch as well) the enormously powerful new neutron and micro-wave radiation sources have meant the opening of fields of research which have previously been inaccessible. The knowledge won here will obviously lead in turn to further progress.

As an example I can mention the use of centimetre waves in the study of the fine structure of spectral lines and thereby in the untangling of coupling conditions within the atom. Further, with the help of the uranium reactor, the study of the ability of various basic elements to capture neutrons during the formation of new radio-active isotopes can be mentioned.

There is much discussion these days about the mutual relationship between "applied research" and "basic research". On an occasion like the present, when a large industrial laboratory is being inaugurated, it could be worthwhile to dwell somewhat on this matter. The differentiation between clearly utilitarian research and seeking the truth for its own sake can also be found in the most primitive forms of research and has no doubt always existed. We have within us a drive to know what makes things tick, as it were - in varying degrees in different individuals - which leads to "useless" truth-seeking. This drive is doubtless very basic: after all, we find it clearly expressed in children and in apes. The drive to find tools to improve one´s living conditions is also basic. Their combination, however, is probably of a much later date and constitutes one of science´s most important resources. New points of departure can seldom be gained by systematically testing all different possibilities. For this, intuitive, unconditional research is needed. The most important discoveries are those which are the least expected and which therefore cannot be pursued directly because we cannot know what we are looking for. When we hear of a new discovery, we are often astonished that nobody had ever thought or experimented along those lines before. The ability to see the unexpected is a rare gift.

SCIENTISTS´ CURIOSITY AT PHENOMENA - A SOURCE OF TECHNICAL DEVELOPMENT

New viewpoints about phenomena surrounding us have often been put forward by researchers who are equipped with thought processes which deviate from those of the majority. "The most important thing for a scientist is to be able to be surprised by the apparently obvious", as the famous chemist Wilhelm Ostwald once said. The classic example is, of course, Newton´s surprise at the falling motion of bodies, which led to the discovery of the laws of gravity (as the poet Kjellgren put it, "You who in the apple´s fall found laws for stars"). Einstein´s ruminations about the enigma of weight and mass led to the discovery of the equivalence of energy and mass and thereby paved the way for the exploitation of atomic energy.

The Indian Raman´s discovery of the spectral effect which bears his name, and which is of such great importance to the study of the structure of molecules, can be traced back to

---

*This lecture by Professor The Svedberg was given in Swedish as an address "Forskningens mål och medel" at the inauguration of the Research Laboratory, Höganäs-Billesholms AB, Sweden, September 10, 1947. It has previously been published in IVA **18**, 127-133 (1947) and is translated by Dr. Donald S. MacQueen, Uppsala University.

157

his amazement at the wonderful blue opalescence of the Mediterranean, when he experienced
that much-celebrated sea for the first time on a trip to Europe. The experimental tools he
used to photograph light diffused by different elements were not new. Any spectroscopist
with normal laboratory equipment would have been able to make Raman's discovery, but nobody
hit upon the idea of setting up just that experiment.

Take another example. Toward the end of the 19th century a series of chemical
discoveries of the most astonishing kind were made in England. It was found that, besides
nitrogen and oxygen and some carbon dioxide and water vapour, the atmosphere contains no
fewer than five hitherto overlooked basic elements. The remarkable thing is that one of
these gases constitutes a full 1.3 per cent of the air, so that on our earth there is
68,510,000,000,000 tonnes of an element which had completely or almost completely eluded the
notice of scientists until the year 1894. I say almost completely, because the ingenious
English chemist Cavendish in the late 1700's made a few observations which suggested that a
then unknown gas was present in air. As he failed to pursue his experiments further, they
remained unnoticed. It was not until 1894, when Lord Rayleigh showed by careful measurements that the nitrogen produced from chemical compounds was somewhat lighter than that
present in the air, that the search for new gases in the atmosphere really got underway.
From this clue, work was carried out first by Rayleigh and the chemist Ramsay and then by
Ramsay alone. The result was the discovery of the inert gases helium, neon, argon, krypton,
and xenon in the atmosphere.

We find an example of how an observer with an open mind for new phenomena can make an
important observation from the wrong starting point in Becquerel's discovery of the radioactivity of uranium. He proceeded from Röntgen's discovery of a penetrating new radiation
which proved to come from the fluorescent glass wall of a cathode steel tube. Becquerel
related this fluorescence – incorrectly – to the production of the new rays and looked for
such radiation from various fluorescent and phosphorescent substances, which were
illuminated by ordinary light. He found an effect from uranium salts and was thorough and
unprejudiced enough to test whether illumination was really necessary. It turned out that
the uranium compounds emitted their radiation regardless of previous illumination, and the
tremendously important discovery of radioactivity was thereby made.

A discovery can often be delayed by the fact that even though the scientist has the
phenomenon under observation, he misunderstands it because of its new, idiosyncratic nature
and promptly proceeds to place it in some familiar category. We have an example of this in
the discovery of the neutron, the elementary particle which has proved to play such an
important role in the structure of matter, and which mediates the chain reaction that makes
the atom bomb explode. The Germans Bothe and Becker found that beryllium, when bombarded by
α-particles, emits radiation of a highly penetrating nature. Since the most penetrant
radiation then known was γ-radiation from radio-active elements, the two scientists drew
the false conclusion that these beryllium rays were a sort of γ-radiation. Irene and
Frederic Joliot-Curie continued the experiment and came across the unexpected phenomenon
that rapid protons were emitted from hydrogen-containing substances which were exposed to
beryllium radiation. Such effects had never been observed in experiments with γ-rays.
Despite this, they stuck to the γ-ray hypothesis. The Englishman Chadwick had the privilege
of solving the riddle. He had worked with Rutherford at the Cavendish Laboratory, where
they had been toying with the idea that there might be such a thing as neutral particles.
Chadwick tested that hypothesis and found that beryllium radiation consisted of a stream of
rapid neutrons. Their mass is nearly the same as that of protons, which explains the
tendency of neutron radiation to cause proton radiation by collision, which had indeed been
observed by Joliot-Curie.

FUNDAMENTAL AND APPLIED RESEARCH MUST BE CARRIED ON SIMULTANEOUSLY

Just as it is important that basic research be given an opportunity to develop freely and
that talented, original minds with their deviant, "crazy" ways of thinking not be stifled,
it is important that organized, goal-directed research be pursued. Its task is to take
those apparently useless ideas and make them useful and to carry out technical and
scientific developmental work. Here group research is obviously indispensable if results
are to be achieved within a reasonable time. But even in the field of basic research it is
necessary to set up group projects to solve problems which require a large array of
experimental and theoretical resources. The time interval between scientific discovery and
technical exploitation has shrunk more and more. In the war effort's combination of basic
research and application, this time difference was reduced to an absolute minimum. We have
every reason to learn a lesson from this, even though the hectic work tempo that the war
forced upon the belligerent countries should no longer prevail.

As an example of technical progress through deliberate research starting with a
"useless" scientific discovery, let me mention the exploitation of inert gases. Their
chemical indifference renders them suitable for filling electrical incandescent light bulbs
in order to prevent the rapid blackening of the lamp by the evaporation of the metal. Inert
gases are also important in another way within the field of illumination technology.
Production of light by glowing solid substances is always extremely uneconomical, even in
improved lamps.
Discharges in rarefied gases convert electric energy more economically into light.

Similar light phenomena were known even to early experimenters (the electric egg, Geissler tubes), but only through technical developmental research have they come to be of practical importance. This is where inert gases are indispensable. For advertising signs we have the familiar neon tubes and for traffic lighting, etc., sodium and mercury lamps. The latter contain sodium vapour and mercury vapour, respectively, with argon as a basic gaseous vehicle. Recently lamps on discharges in mercury vapour have begun to play an ever greater role. To start with, the vapour pressure was kept high (several atmospheres). Such bulbs yield strong, blue-green light. To make it whiter, bulbs are built with a combination of metal filaments and mercury, and their use is spreading for home lighting as well. The latest step forward consists of having an electric current work upon highly diluted mercury vapour, which thereupon mainly emits ultra-violet rays, and then converting this ultra-violet light into visible light with the help of fluorescent substances placed on the inner wall of the lamp. This is a fresh example of how technology is connected to old laboratory experiments. As a prototype for these fluorescent or phosphorescent substances, we can in fact go back to Lapis Bologniensis, which was discovered by the cobbler Vincentius Casciarolus in the early 1600s and which later, like so many other substances of this type, was the object of thorough studies in physics and chemistry laboratories (by Becquerel, the father of radio-activity, to name one). Fluorescent lamps give off a light of the same character as that of daylight under a slightly cloudy sky, and the energy they save is about 65 per cent in relation to the incandescent bulb. We are onto a line of development which no doubt will dominate illumination technology for a long time to come. It is worth mentioning, by the way, that these modern discharge lamps could not have been created if an important aid, the oxide filament glow cathode, had not been transplanted into technology from the physics laboratories.

The steel industry is indebted for its rapid development to modern metallography and metallurgy, which have made possible a superb mastery of that material through alloying additives. Here basic research, in the most intimate cooperation with technology, has scored one of its greatest triumphs. Swedish research once dominated metal chemistry, and we are still on the cutting edge. Of the seven most prominent steel processing metals, no fewer than four were first produced in Sweden: nickel by Cronstedt in 1751, molybdenum by Hjelm in 1782, cobalt by Brandt in 1742, vanadium by Sefström in 1830 (while manganese was discovered in Germany by Gahn in 1774, chromium in France by Vauquelin in 1798, and wolfram ´tungsten´ in Spain by d´Elhuyar in 1783).

Non-metallic materials for all sorts of utensils and for building purposes were formerly taken straight from nature, and consequently they were but little adaptable. In this field basic research, in cooperation with the chemical industry, has accomplished much. This goes for inorganic (ceramic) materials as well as organic-synthetic ones. In some cases, for example, polystyrene, bakelite, and melamine, the basic observations were carried out in scientific laboratories a hundred years ago, but the discoveries went long unnoticed. The synthetic rubbers Buna and Neoprene, on the other hand, are the result of deliberately combining basic research and technical developmental work. In the fields of both inorganic and organic materials, new experimental aids have played a major part. We need only recall thermal analysis, x-ray analysis, sedimentation analysis, spectral analysis, the polarization microscope, the electron microscope, and how most recently the radioactive indicator method – to mention a few research tools in the inorganic field. For organic materials we have infra-red spectroscopy, Raman spectroscopy, mass spectroscopy, elasticity measurement methods, ultra-centrifugation, diffusion measurement, electrophoresis, flow-birefringence, adsorption analysis, dipole measurement, ultrasound, and here too the indicator method with radio-isotopes and heavy isotopes as tracer elements.

THE UTILISATION OF ATOMIC ENERGY – THE GREATEST COOPERATIVE RESEARCH ACHIEVEMENT IN THE HISTORY OF MANKIND

The greatest cooperative research accomplishment in the history of mankind is the finding of means to exploit atomic energy. Rutherford´s splitting of the nucleus of the nitrogen atom in 1919, with the help of α-particles from radio-active substances is the starting point. There then followed a series of "useless" laboratory experiments, among which we can mention Cockcroft´s and Walton´s splitting of atoms with artificial radiation sources, Chadwick´s discovery (mentioned earlier) of the most reactive of all particles, the neutron in 1932, plus Fermi´s demonstration of its ability easily to penetrate atomic nuclei whose equilibrium is being disturbed. Twenty years after the first artificial splitting of the atom, Hahn discovered in 1939 an entirely new type of atomic reaction, so-called nuclear splitting or fission. This entails the releasing of tremendous amounts of energy, and by being set up as a chain reaction with or without speed regulators, it can serve either peaceful power production in so-called uranium piles (or reactors) or martial destruction in the atom bomb.

The astounding progress in our knowledge of atomic energy, which has been made during the war, especially in America, has spurred us onward to great efforts to continue work in this area. Cyclotrons of great dimensions are being built in several places in the U.S.A. and Canada. Besides the giant cyclotron in Berkeley, which was started before the war and is now complete, and which has a pole diameter of 184", there will soon be a 130" cyclotron in Rochester and one with a 118" pole diameter at Harvard University. At the east coast´s

great atomic research centre in Brookhaven, Long Island, outside New York, there are plans
for not only an "ordinary" 60" cyclotron for 30 MeV protons but also a frequency modulated
240" cyclotron for 600 MeV protons. With the help of these protons, they hope to be able to
produce mesons, those enigmatic particles of cosmic radiation which are also probably found
in atomic nuclei as a sort of immaterial cement between protons and neutrons. Finally work
is underway in Brookhaven on a giant machine of the synchrotron type designed to yield
10,000 MeV (10 BeV) protons. It is designed to produce pairs of protons (consisting of one
regular positive and one negative proton). The magnet for this apparatus will have a radius
of 160´ with ring-shaped pole shoes 4´ in breadth, and its weight is estimated to be 17,000
tonnes.

     To the best of our knowledge, America is far ahead of other countries in the field of
atomic research. Here in Sweden we can obviously by no means keep up with developments in
the U.S.A. Atomic energy research has, however, taken rapid strides even here at home,
thanks to a lively interest on the part of researchers and substantial support from the
state and - in one case at least - from industry. Two large cyclotrons are under
construction, one for Professor Manne Siegbahn in Stockholm and one at the Institute of
Physical Chemistry in Uppsala. The latter, which will have a pole diameter of 90" and with
the help of which we hope to be able to produce 120 MeV protons, will weigh 640 tonnes.
It is being funded for the most part by the Werner Concern in Gothenburg and will be used,
among other things, for the technical study of radiation-related chemical reactions.

METHODS OF MEASUREMENT OFTEN UNIVERSAL

Laboratories of the past were strictly specialized, in that a medical laboratory had its own
methods, an organic chemistry laboratory others, inorganic its own, and physics still
others. Nowadays the tools of research are tending more and more to be common to all
branches of experimental natural science and medicine. It is often difficult to tell at a
glance which science or industry a laboratory is designed to serve. Of course, the various
branches of research require many special aids, indispensable to the solution of details,
but the general equipment and instruments are more or less universal. So too are the
theoretical outlooks growing more generally relevant with every new year. A physiologist, a
chemist, a physicist must know the same general fundamentals of the behaviour of energy and
matter. Specialisation is necessary, of course, but the broad view is no less important.
The formerly so honorable tile of "specialist" is nowadays a rather dubious compliment!

     The steadily growing arsenal of aids necessitates cooperation and still more
cooperation if results are to be achieved quickly. It is impossible for the individual
scientist fully to master the whole fields of knowledge and proficiency that research
represents. In America in particular it is now common to have work teams composed of
mathematicians, physicists, technologists, chemists, medical researchers, etc. Publications
of research results are often signed by a dozen or more names. Objections have been raised
against group research here in Sweden on the grounds that Swedes are poorly suited to such a
form of work: they are individualists and prefer to work alone. My experience from
attempts at organizing teamwork within a scientific institution has been, on the contrary,
highly encouraging. Individuality can very easily find expression and thrive in work toward
a common goal.

     The state should obviously assume the main responsibility for the costs of basic
research, but industry´s research too ought to touch upon basic research in order to be
fruitful and not run the risk of losing touch with basic developments. The scientist´s
collaboration in industry, to be effective, should not be limited to occasional assignments
and consultations when something goes wrong, but rather should be such that a continuous
joint effort between basic research, applied research, and industrial production is
achieved.

EDUCATION FOR RESPONSIBILITY THE BEST GUARANTEE AGAINST MISUSE OF SCIENTIFIC RESOURCES

At the outset of my talk here today I touched upon the risk that arises when even more
powerful material aids are placed in our hands without guarantees being given that prevent
their misuse. War is nowadays entirely based on the belligerent parties´ trying to outdo
one another in the production of new offensive and defensive means. Scientific and
technological research plays a dominant role here. The question has been asked in many
places in this post-war period what scientists and technologists should do in the event of
the threat of war. Should they go on strike to prevent the breaking out of war? Such a
general strike against war on the part of research must be total, however, to be effective.
Is it reasonable to expect that such unity could be reached at the present time? Hardly! We
must proceed along the path of enlightenment and education and seek to bring about a deep
change of heart, a will to negotiate and cooperate between people and nations. Should we
then put an end to scientific and technological research until we are more worthy of its
results? Such a thing would be impossible, even if we were to find it desirable. Instead,
we must hasten by all possible means the growth of responsibility. A couple of articles in
the English magazine Nature dealt with these problems some half a year after the end of the
World War. There the following was expressed. "It cannot possibly be a matter of
discontinuing scientific studies until mankind is better suited to receive their results. If

research in the material field has passed by our progress concerning our knowledge of man, then the tempo in the study of man as a social being must be stepped up until the two can walk side by side, carrying mankind forward toward the higher ideals of life for which the best of every generation have always striven". The difficult question is how this tempo in the study of man can be hastened. The methods there are not as clear as those used in natural science and medicine. And I am sure few would consider possible the application of a similar methodology within psychology, sociology, and political science. Intensive activity for enlightenment in these areas must sooner or later bear fruit. If we become clearly aware of the nature and causes of our affects, we ought to be able to control them more easily. But, as we all know, it is no simple task we have before us. "It is easier for an atom bomb to wipe out a city than a complex of the soul", someone recently said. Every nation as well as every individual must learn to curb his urge to demand for himself without consideration for others, must be freed from the constant fear of losing what he possesses. Such a change of heart can only be achieved through deep insight into the causes of our spiritual complexes, through understanding natural phenomena, within as well as outside ourselves, and through insight about the greater whole. We are lost if we cannot exploit the tools of research for the common good and learn to strive toward common goals.